AF301198

FSC
www.fsc.org
MIX
Papier aus verantwortungsvollen Quellen
Paper from responsible sources
FSC® C105338

Mebrahtu Haile

Biofuel Energy

spent coffee grounds biodiesel, bioethanol and solid fuel

Anchor Academic
Publishing

Haile, Mebrahtu: Biofuel Energy : spent coffee grounds biodiesel, bioethanol and solid fuel, Hamburg, Anchor Academic Publishing 2014

Buch-ISBN: 978-3-95489-305-8
PDF-eBook-ISBN: 978-3-95489-805-3
Druck/Herstellung: Anchor Academic Publishing, Hamburg, 2014

Bibliografische Information der Deutschen Nationalbibliothek:
Die Deutsche Nationalbibliothek verzeichnet diese Publikation in der Deutschen Nationalbibliografie; detaillierte bibliografische Daten sind im Internet über http://dnb.d-nb.de abrufbar.

Bibliographical Information of the German National Library:
The German National Library lists this publication in the German National Bibliography. Detailed bibliographic data can be found at: http://dnb.d-nb.de

Summary

In this study, the utilization of waste coffee residue for biodiesel production, its solid byproduct after oil extraction for bioethanol production, as well as the second byproduct after bioethanol production for solid fuel and compost production was investigated. For the study, waste coffee residue sample was collected from TOMOCA PLC, Addis Ababa, Ethiopia. The oil was then extracted using n-hexane and resulted in oil yield of 19.73 %w/w. The biodiesel was obtained by a two-step process, *i.e.* acid catalyzed esterification followed by base catalyzed transesterification using catalysts sulfuric acid and sodium hydroxide respectively. The conversion, after esterification of waste coffee residue oil in to biodiesel, was about 80.4%. Various parameters that are essential for biodiesel quality were evaluated using the American Standard for Testing Material (ASTM D 6751- 09). The results obtained for kinematic viscosity (5.3 mm²/s), carbon residue (0.033%), flash point (222°C), ash content (0.0123%), water and sediment (<0.01%), iodine value (73.41 gI2/100g), acid value (0.78), copper striping corrosion (1a) and calorific value (38.4 MJ/kg) revealed that all quality parameters are within the range specified except for acid value. The fatty acid composition of the biodiesel was also analyzed by Gas chromatography and the major fatty acids found were linoleic acid (39.8%), palmitic acid (37.6%), oleic (12.7%), and stearic acid (7.6%). In addition, the solid waste remaining after oil extraction was investigated for possible use as a feedstock for the production of bioethanol. Hydrolysis of the spent was carried out using dilute sulfuric acid followed by fermentation using *S. cereviciae*, and resulted in bioethanol yield of 8.3 %v/v. Furthermore, the solid waste remaining after bioethanol production was evaluated for compost and solid fuel applications. The result indicated that the processed coffee residues could still be used as compost (21.9:1 C/N) and solid fuel (20.8 MJ/Kg). Therefore, the results of this work may suggest a new insight to production of biofuel from waste materials.

Keywords: Waste coffee residue, Biodiesel, spent of WCR, Saccharomyces cereviciae, Bioethanol, solid fuel, compost

Acknowledgement

First and foremost I would like to acknowledge gratefully to my research advisors Dr. Araya Asfaw and Dr. Nigist Asfaw for their generous and continuous support, encouragement and advices given to me while doing this research. Their invaluable ideas and help made my work a lot easier.

I would like also to acknowledge Dr. Mesfin Redi for his invaluable ideas and excellent guidance given to me while doing this research. Not forgetting his excellent supervision which brought me to successfully complete this research.

I am gratefully acknowledging Horn of Africa-Regional Environment Center/Network (HoAREC/N) for the financial support that made this study possible. In addition, I would also like to thank all the staffs of the organization for their grateful cooperation. Next, I would like to express my gratitude to AAU, department of Chemistry, especially the technicians and laboratory assistants for their technical support and assistance. In addition, my special thanks to W/t Senayt Dangne and Mr. Kehali Kiros for their technical guidance and assistance.

I would like to express my best acknowledgment to Mr. Akalu Wube the manager of TO.MO.CA PLC, Ethiopian petroleum supply enterprise (EPSE), Ethiopian Health and Nutrition Research Institute (EHNRI), Ethiopian Geological survey laboratories and Ethiopian ministry of mines and energy. I am especially thankful to Mr. Mengesha and Mr.Agignichew from EPSE and Mr. Abel W/tinsae and Mr. Daniel Abera from EHNRI and other staff members for their professional guidance and constant encouragement up to the end of the lab work.

Last but not least, I would like to thank my family for their continuous love, sacrifice and supports given to me while I am on this work.

Table of Contents

List of Figures

List of equations

Acronyms

ASTM	American Society for Testing and Materials
AV	Acid Value
BD	Biodiesel
CN	Cetane Number
CP	Cloud Point
CV	Calorific Value
EN 14214	European Standards for Biodiesel
EPA	Environmental Protection Agency
FAME	Fatty Acid Methyl Ester
FBP	Final Boiling Point
FA	Fatty Acid
FFA	Free Fatty Acids
FTIR	Flourier Transform Infrared Spectroscopy
GC	Gas Chromatography
GHG	Green House Gases
IBP	Initial Boiling Point
IV	Iodine Value
KV	Kinematic Viscosity
LPG	Liquid Petroleum Gas
MoME	Ministry of Mines and Energy
PP	Pour Point
Rpm	Revolution per minute
S. cerevisiea	*Saccharomyces cerevisiae*
SG	Specific Gravity,
ULSD	Ultra Low Sulphur Diesel
WCR	Waste Coffee Residue

1. INTRODUCTION

1.1.   Background and justification

In recent times, the world has confronted with crises of increased demand for energy, price hike of crude oil, global warming due to emission of green house gases, environmental pollution, and fast diminishing supply of fossil fuels (Atadashi, *et al.*, 2011; Miguel and Calixto, 2009). The indiscriminate exploration and consumption of fossil fuels has led to a reduction in petroleum reserves (Miguel and Calixto, 2009).

Our reliance on these energy sources threatens energy security and influence economic growth especially in fuel importing countries like Ethiopia. About all of Ethiopia's liquid fuel requirements are imported in the form of refined petroleum products (Alemayhu, 2007). This external energy supply is unsteady and has become a burden to the rapidly growing national economy. In addition, diesel powered motor vehicles in the road transport sector are an important contributor to the total gas emissions in the urban cities (Christoffel, 2010). From the point of view of global environment protection and the concern for long-term supplies of conventional diesel fuels, it becomes necessary to develop alternative fuels comparable with conventional fuels. Alternative fuels should be, not only sustainable but also environmentally friendly (Miguel and Calixto, 2009). Some of the most notable alternative sources of energy capable of replacing fuels (Miguel and Calixto, 2009) include amongst others: water, solar and wind energy, and biofuels (Atadashi *et al.*, 2011). A potential diesel oil substitute is biodiesel (Miguel and Calixto, 2009).

Biodiesel is a new energy source that has grown in importance over recent years. Nowadays, used vegetable oils are potential renewable sources for the production of biodiesel as an alternative to petroleum based diesel fuel, which is derived from diminishing petroleum reserves and which has environmental consequences caused by the exhaust gases from diesel engines (Maceiras *et al.*, 2009). Biodiesel has several benefits such as a diminution in greenhouse gas emissions: it reduces emissions of carbon monoxide by about 50% and emissions of carbon dioxide by about 78% (Sheehan *et al.*, 2008). In addition, biodiesel is produced from a variety of vegetable oils (such as soybean, rapeseed and sunflower) and animal fats, and can be used in diesel engines blended with petroleum diesel or on its own (Sánchez *et al.*, 2012). Researchers are developing certain crops with high oil content just for the production of biodiesel (Azam *et al.*, 2005; Cardone *et al.*, 2003; Encinar *et al.*, 2002; Gressel, 2008) or looking for new sources to produce biodiesel (Kondamudi *et al.*, 2008). Therefore, it will be very useful to look for new raw sustainable materials for biodiesel production that do not involve the use of cereals and plants that compete with land.

Coffee is one of the largest agricultural products that are mainly used for beverages (Kondamudi *et al.*, 2008) throughout the world and providing approximately 30.6% of Ethiopia's foreign exchange earnings in 2010-2011 (Bureau of African Affairs, 2012). Ethiopia is currently producing an estimated 9.804 million 60-kg bags that would rank as the third largest coffee producer in the world after Brazil and Vietnam (African Commodity Report, 2012) and half of the coffee is consumed by Ethiopians (Abu, 2012).

Waste coffee residue (WCR), the solid dregs found from the treatment of coffee powder with hot water to prepare instant coffee, is the main coffee industry residues with a generation of 6 million tons worldwide (Tokimoto *et al.*, 2005) and 235,296 tons in Ethiopia. According to Kondamudi *et al.* (2008), on a worldwide scale, based on the amount of coffee that is used, 340 million gallons of biodiesel can be produced from Waste coffee residues. Simões (2009) demonstrated that WCR can be used for the production of biodiesel and fuel pellets and as a source of polysaccharide with immune stimulatory activity. The amount of oil in the Waste coffee residue source varies from 11 to 20 wt % depending on its types (Arabica and Robusta) (Daglia *et al.*, 2008), which is roughly equivalent to that of palm, rapeseed, and soybean sources (Kondamudi *et al.*, 2008). Compared to other waste sources; such as cooking oils, animal fats, and other biomass residues, coffee has the additional benefits of being less expensive, more stable due to the presence of antioxidants, and comprising of a nice smell (Al-Hamamre *et al.*, 2012). On average, the Waste coffee residues comprise 15% oil, by weight, which can be converted to a similar amount of biodiesel using transesterification methods (Kondamudi *et al.*, 2008).

The biodiesel from waste coffee residues possesses better stability than biodiesel from other sources due to its high antioxidant content (which hinders the rancimat process) (Campo *et al.*, 2007; Yanagimoto *et al.*, 2004). WCR is also considered an inexpensive and easily available adsorbent for the removal of cationic dyes in wastewater treatments (Franca et al., 2009). However, none of these strategies have yet been routinely implemented, and most of these residues remain unutilized, being discharged to the environment where they cause severe contamination and environmental pollution problems due to the toxic nature (presence of caffeine, tannins, and polyphenols) (Leifa *et al.*, 2000). Nowadays, there is great political and social pressure to reduce the pollution arising from industrial activities. In this sense, conversion of WCR to value-added compounds is of environmental and economical interest (Mussatto *et al.*, 2011). In this work, investigation of WCR as a potential raw material for biodiesel production and its by-products for bioethanol and solid fuel as well as compost was carried out.

1.2.   Problem of the statement

In today's world, alternative fuels are needed more than ever. The primary sources of energy are mainly non-renewable: natural gas, oil, coal, peat, and conventional nuclear power. These Conventional fuels are constantly being depleted; however, our dependency on these fuels is still growing. Additionally, the price on foreign fuels is ever increasing. For these reasons, Ethiopia is pursuing alternative fuel sources to lessen the dependency on conventional fuel that is petro. One alternative fuel is biodiesel; biodiesel can be produced from vegetable oil or animal fat and thus can be used to alleviate the foreign fuel dependency. In order for biodiesel to be a viable alternative fuel source, a cheap feedstock for biodiesel production process needs to be improved. Compared to current designs and fossil fuel, the process must be cost competitive.

Ethiopia imports its almost all petroleum fuel requirement and the demand for petroleum fuel is rising rapidly due to a growing economy and expanding infrastructure (Alemayhu, 2007). Statistics from the Ethiopian Ministry of Mines and Energy (MoME) indicate that the country spends about Ethiopian Birr 10 billion (US$800 million) annually to import petroleum products for domestic consumption. This astounding figure represents nearly 90 percent of the earnings that the country makes each year in foreign trade. By cutting its dependency on foreign oil, Ethiopia could perhaps keep some of the money inside the country (Gathanju, 2010).

1.3. Objectives

1.3.1. General objective

- To Investigate the Waste Coffee Residue (WCR) remaining after brewing coffee as a potential alternative raw material for biodiesel production.

1.3.2. Specific objectives

- To extract oil from WCR remaining after brewing coffee
- To analyze Physicochemical parameters of WCR oil that affect the production and characteristics of the biodiesel produced from the oil
- To produce biodiesel from WCR and to analyze its fuel properties such as acid value, density, kinematic viscosity, iodine value, flash point, cetane number, carbon residue, water and sediment content, heating value and cloud point, along with the standard test methods
- To determine fatty acid composition of the biodiesel
- To evaluate the spent remaining after the oil extraction for possible uses as fuel and compost

1.4. Significance of the study

The results of this study will give insight to produce biodiesel from waste materials. Clean energy for today's economic development is crucial to assure a sustainable development. Hence, the investigation of waste coffee residue remaining after brewing coffee as a potential alternative raw material for biodiesel production is very timely because of arising problems such as the rising cost of fuel in the market, global warming phenomenon, and health problems such as respiratory diseases caused by the harmful byproducts of burning petroleum-based fuels. Producing a biodiesel from it will be a good way to minimize GHG and waste in the environment.

2. LITERATURE REVIEW

2.1. Overview of Ethiopia's Energy Sector

Ethiopia is well endowed with a variety of energy and other natural resources. However, much of the energy resource available has yet to be exploited. The renewable energy resources with potential include biomass, hydropower, and alternative forms of energy-solar, wind and geothermal energy. There are also considerable reserves of coal and natural gas (W/Giorgis, 2004).

Most people living in Ethiopia have, until now, been unable to satisfy their household energy requirements with modern energy sources (kerosene, electricity, gas). In rural areas they use only biomass (wood, dung or agricultural waste) for cooking, baking and heating. The biomass energy consumption is estimated about 94% of the total. The low agricultural production is a consequence of deforestation, erosion and desertification. On the other hand petroleum products which are used mainly at urban household center are entirely an imported commodity. Demand for these products is rising rapidly increasing due to in scarcity of fuel wood and the change of life style of the people. The rise in demand is accompanied by a much faster growth in the import bill because of rising petroleum prices and products (Alemayhu, 2007).

In Ethiopia the gross available potential land for production of feedstock for biodiesel is estimated about 23,305,890 hectares and the total irrigable land for sugarcane production for ethanol production is about 700,000 hectares. Thus Ethiopia has a potential to produce 1billion liters of ethanol within available suitable land (Alemayhu, 2007). The major use of energy, about 89% of the overall energy consumption in the country, is the households. The second most important sector in terms of energy consumption is industry (4.5%) followed by services and others (3.6%) while agriculture and transport were attributed to the remaining 2.3%. The consumption of energy is directly related to the availability of energy source, the size of the population and the price (Ministry of mines and energy, 2011).Table 2.1 shows the sector-wise percentage usage distribution of energy source type in Ethiopia.

Table 2.1: Sector wise energy source utilization percentage distribution

Sectors	Energy source		
	Biomass (%)	Petroleum (%)	Electricity (%)
Households	98.6	1.1	0.3
Industry	75.7	17.3	7
Services	94.3	1.3	4.4
Transport	-	100	-
Agriculture	-	100	-

Source: (Meskir, 2007; Ministry of mines and energy, 2011)

2.2. Introduction to biodiesel

Biodiesel is defined as "a substitute for, or an additive to diesel fuel that is derived from the oils and fats of plants and animals" (Ma and Hanna, 1999) or a fuel composed of monoalkyl esters of long-chain fatty acids (Abreu *et al.*, 2004; Hass *et al.*, 2001) produced by transesterification reaction of vegetable oils and animal fats with short chain alcohols meeting the requirements of ASTM D6751 (ASTM, 2008; Burtis,2006; Encinar *et al.*, 2002; Noureddini *et al.*,1998). The alcohol and the base compound (lye) are used to split vegetable oil into components. The ester component is what is known as biodiesel, while the separated by-product is glycerin (Alberta, 2006). Glycerin has value as a co-product and can be used in soaps, lotions, and as a lubricant (Sawyer, 2007). Biodiesel is a form of biofuel that is a nontoxic, biodegradable substance and does not contain any sulfur (Alberta, 2006). It is an alternative to diesel fuel produced from domestic renewable resources. Typically biodiesel derived from the oils and fats of that of sunflowers, soybean, canola, and rapeseeds, can be used in diesel engines with no to little modifications (Liu *et al.*, 2010).

Similar to diesel, biodiesel can function as an alternative to electricity and petroleum diesel. Because its components are derived from renewable resource and has a lower emission compared to the traditional petroleum fuel, biodiesel is environment-friendly. With every one unit of needed energy to generate biodiesel, in turn, 4.5 units of energy is return. This is due to biodiesel having a high-energy balance and is locally produced (National Biodiesel Board, 2010).

2.3. Biodiesel production

Bio-diesel production is a very modern and technological area for researchers due to the relevance that it is winning everyday because of the increase in the petroleum price and the environmental advantages (Marchetti *et al.*, 2005). The most common method of producing bio-diesel is the reaction of vegetable oils and animal fats with alcohol mainly methanol in the presence of catalyst. This process is called transesterification and is not a new process. It was conducted as early as 1853 by two scientists E. Duffy and J. Patrick. Since that time several studies have been carried out using different oils such as soybean (Watanabe *et al.*, 2002), rapeseed (Kusdiana and Saka, 2001), cotton seed (Royon *et al.*2007), waste cooking (Zhang *et al.*, 2003), spent coffee ground (Daglia *el al.*, 2008; Kondamudi *et al.*, 2008),sunflower seed (Harrington and D'Arcy-Evans,1985) , winter rape (Peterson *et al.*, 1991), different alcohols such as methanol (Demirbas, 2006), ethanol (Encinar *et al.*, 2002).

2.3.1. The Transesterification (alcoholysis) Process of Biodiesel

The vegetable oils and animal fats usually contain free fatty acids, phospholipids, sterols, high viscosity, water, odourants and other impurities. Because of these, direct use of these vegetables and animal fats oil as fuel for diesel engine can cause particle agglomeration, injector fouling due to its low volatility and high viscosity, which is about 10 to 20 times greater than petroleum diesel (Fan, 2008). To reduce these problems the oil requires chemical modification principally through transesterification, pyrolysis and emulsification (Ma and Hanna, 1999). Among these, the transesterification is the main and fore most important step to produce the cleaner and environmentally safe fuel from vegetable oils (Knothe and van Gerpen, 2005; Meher *et al.*, 2004). Additionally, the physical properties of biodiesel produced by this simple process are very close to the petroleum diesel fuel (Fan, 2008).

Transesterification is the general term used to describe the important class of organic reactions where an ester is transformed into another through interchange of the alkoxy moiety (Freedman *et al.*, 1986). It is the reaction of vegetable oil or animal fat with an alcohol, in most cases methanol, to form esters and glycerol. According to Srivastava & Prasad (2000) transesterification is the displacement of alcohol from an ester by another alcohol in a process similar to hydrolysis, except that alcohol is employed instead of water. The transesterification reaction is affected by alcohol type, molar ratio of glycerides to alcohol, type and amount of catalyst, reaction temperature, reaction time and free fatty acids and water content of vegetable oils or animal fats. The transesterification process consists of a sequence of three consecutive reversible reactions, which include conversion of triglycerides to diglycerides, followed by the conversion of diglycerides to monoglycerides. The glycerides are converted into glycerol and yield one ester molecule in each step.

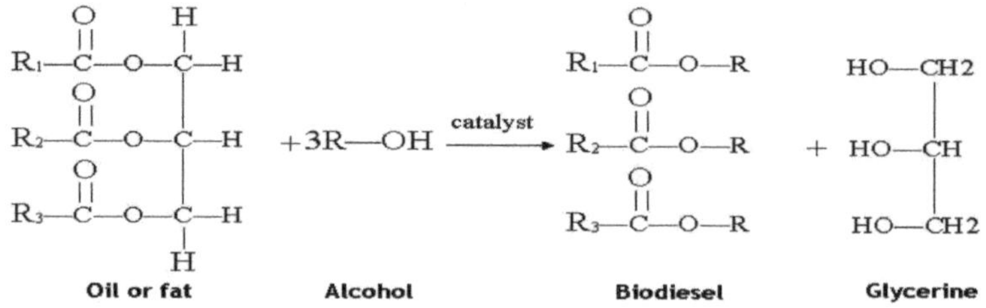

$$
\begin{array}{c}
\text{Oil or fat} + 3R\text{—OH} \xrightarrow{\text{catalyst}} \text{Biodiesel} + \text{Glycerine}
\end{array}
$$

Fig. 1: Scheme of general transesterification reaction (source: Ma and Hanna, 1999).

2.3.1.1. Catalytic transesterification

Catalytic transesterification is the process by which different catalysts are used to initiate the esterification for making biodiesel. Also known as methanolysis, this process is well studied and established (Helwani *et al.* 2009). The three basic Catalysts of biodiesel production from oils/fats are the base-catalyzed transesterification, the acid catalyzed esterification, and enzymatic catalysis (Kaieda *et al.*, 1999; Haas *et al.*, 2006; Ma and Hanna, 1999; Meher *et al.*, 2006). Among these base-catalyzed transesterification is involved in biodiesel production today (Srivastava and Prasad, 2000; Zhang *et al.*, 2003), where feedstocks with a high water or free fatty acid (FFA) content needs pretreatment with an acidic catalyst in order to esterify FFA (Freedman *et al.*, 1984; Kaieda *et al.*, 1999).This is the most common method done because it is the most economical (Singh *et al*, 2006).

2.3.1.1.1. Base-Catalyzed transesterification

Base-catalyzed transesterification involves stripping the glycerin from the fatty acids with a catalyst such as sodium or potassium hydroxide and replacing it with an anhydrous alcohol, usually methanol. The resulting raw product is then centrifuged and washed with water to cleanse it of impurities. This yields methyl or ethyl ester (biodiesel), as well as a smaller amount of glycerol, a valuable by-product used in making soaps, cosmetics and numerous other products (Refaat, 2010).The base-catalyzed transesterification of vegetable oils and fats to form alkyl ester is faster than the acid-catalyzed reaction. It is the one most used technique commercially as it is the most economical process since it requires only low temperatures and pressures; produces over 98 % conversion and involves direct conversion to biodiesel with no intermediate compounds; also, no special materials of construction are needed (Dube *et al.*, 2007; Freedman *et al.*, 1984; Singh *et al.*, 2006). The most commonly used alkali catalysts are NaOH, CH3ONa, and KOH (Korytkowska *et al.*, 2001; Varghaa *et al.*, 2005; Vicente *et al.*, 2004). Alkyl oxide solutions of sodium methoxide or potassium methoxide in methanol, which are now commercially available, are the preferred catalysts for large continuous-flow production processes (Singh *et al.*, 2006).

Base-catalyzed transesterification, however, has some limitations among which are removal of these catalysts is technically difficult and brings extra cost to the final product (Demirbas, 2003; Goff *et al.*, 2004) and it is sensitive to FFA content of the feedstock oils and water (Canakci *et al.*, 2003; Furuta *et al.*, 2004; Leung and Guo, 2006). The starting material should have free acid content less than 0.5% (acid value less than 1) and water content less than 0.3% (Wright *et al.*, 1944). Other disadvantage of the base-catalyzed transesterification is that the process is energy intensive, recovery of glycerol is difficult, alkaline catalyst has to be removed from the product and alkaline waste water requires treatment (Meher *et al.*, 2006). A typical base-catalyzed process for the production of bio-diesel is shown in Fig. 2.

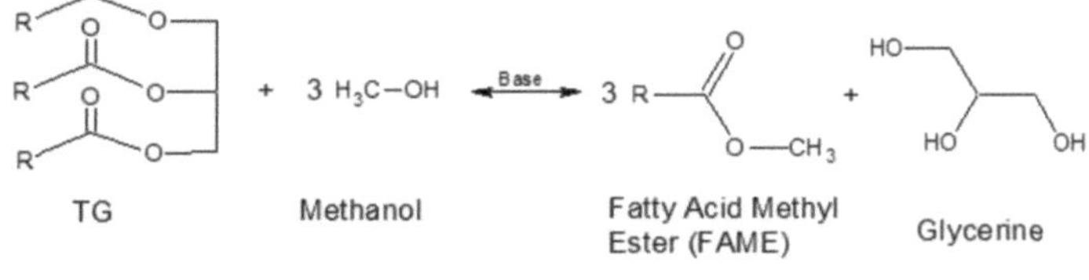

R   R   R   +   3 H₃C—OH   Base   3 R   O—CH₃   +   HO   HO   OH

TG   Methanol   Fatty Acid Methyl Ester (FAME)   Glycerine

Fig. 2: A typical base-catalyzed process for the production of bio-diesel (source: Mohammed, 2011).

2.3.1.1.2. Acid-Catalyzed Transesterification

Acid-catalyzed transesterification is slower than base-catalyzed transesterification. The reaction temperature is usually above 100 °C and reaction time is 3 to 48 hours except when the reaction is conducted under high temperature and pressure (Allen *et al.* 1945; Taylor *et al.* 1927). The procedure of acid-catalyzed transesterification is different from base-catalyzed one. The reaction is refluxed at or near the boiling point of the mixture of cosolvent and alcohol. The transesterification process is catalyzed by sulfuric (Goff *et al.*, 2004; Liu *et al.*, 2006; Lopez *et al.*, 2005), hydrochloric (Lee *et al.*, 2000; Goff *et al.*, 2004), and organic sulfonic acids (Stern and Hillion, 1990). A simplified block flow diagram of the acid-catalyzed process is shown in Fig. 3. In general, acid catalyzed reactions are performed at high alcohol-to-oil molar ratios, low-to moderate temperatures and pressures, and high acid catalyst concentrations. However, ester yields do not proportionally increase with molar ratio. For instance, for soybean methanolysis using sulfuric acid, ester formation sharply improved from 77% using a methanol-to-oil ratio of 3.3:1 to 87.8% with a ratio of 6:1. Higher molar ratios showed only moderate improvement until reaching a maximum value at a 30:1 ratio (98.4%) (Lotero *et al.*, 2006). Despite its insensitivity to free fatty acids in the feedstock, acid-catalyzed transesterification has been largely ignored mainly because of its relatively slower reaction rate (Zhang *et al.*, 2003).

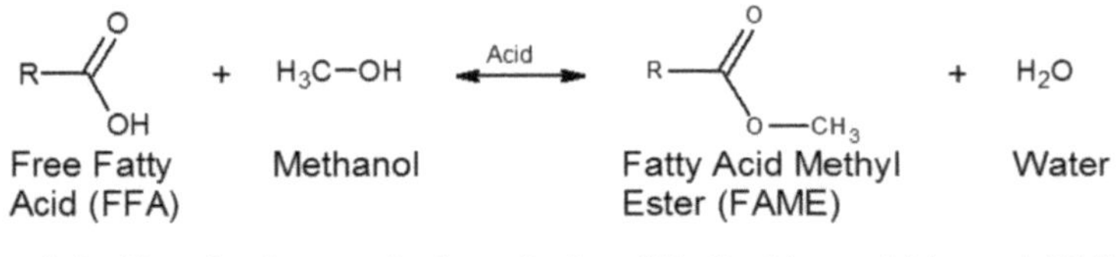

Fig. 3: A typical acid-catalyzed process for the production of bio-diesel (source; Mohammed, 2011).

2.3.1.1.3. Enzyme-Catalyzed Transesterification

Enzyme-Catalyzed Transesterification is gaining more attention nowadays and has the potential to do better than chemical catalysts for biodiesel production in the future. New biochemical routes to biodiesel production, based on the use of enzymes, have become very interesting (Chang *et al.*, 2005; De Oliveira *et al.*, 2004; Noureddini *et al.*, 2005). Most of the articles published have used a variety of substrates such as rice bran oil, canola, sunflower oil, soybean oil, olive oil, and castor oil. Several lipases from microbial strains, including *Candida Antarctica* (Lai *et al.*, 2005; Royon *et al.*, 2007), *Candida rugasa* (Chen and Wu, 2003; Linko *et al.*, 1998), *Pseudomonas cepacia* (Deng *et al.*, 2005; Shah and Gupta; 2007), *Thermomyces lanuginosus* (Xu *et al.*,2004), *Pseudomonas sp.* (Lai *et al.*, 1999), and *Rhizomucor miehei* (Lai *et al.*, 1999; Skagerlind *et al.*,1995) have been reported to have transesterification activity. The enzymatic alcoholysis of soybean oil with methanol and ethanol was investigated using a commercial, *immobilized lipase* (Bernardes *et al.*, 2007; De *et al.*, 2006). In that study, the best conditions were obtained in a solvent-free system with ethanol/oil molar ratio of 3.0, temperature of 50 °C, and enzyme concentration of 7.0% (w/ w). They obtained yield 60% after 1 h of reaction.

The advantages of lipase-catalyzed transesterification, compared to the chemically-catalyzed reaction, are emphasized and can be reused without separation (Nelson *et al.*, 1996; Shimada *et al.*, 2002). Also, the operating temperature of the process is low (50 °C) compared to other techniques (Nelson *et al.*, 1996; Shimada *et al.*, 2002). The main problem of the lipase-catalyzed process is the high cost of the lipases used as catalyst (Royon *et al.*, 2007) and inhibition effects which were observed when methanol was used (Nelson *et al.*, 1996; Shimada et al., 2002).

2.4. Variables Affecting the Transesterification Process

There are number of factors which could affect the transesterification process. These factors include moisture content, free fatty acid contents, molar ratio of oil to alcohol, type and amount of catalyst, reaction time, reaction temperature, mixing intensity, and co-solvent (Demirbas, 2007; Ma & Hanna, 1999; Meher *et al.*, 2006; Sharma *et al.*, 2008). These factors or variables usually have different effect on the transesterification process depending on the method used for the transesterification process. The effects of these factors are described below.

2.4.1. Effect of free fatty acids (FFA) and moisture

The free fatty acids (FFA) and moisture contents are two key parameters for determining the viability of the feedstock (vegetable oil) transesterification process (Meher *et al.*, 2006). In the transesterification, FFAs and water always produce negative effects, since the presence of FFAs and water causes soap formation, consumes catalyst and reduces catalyst effectiveness, all of which result in a low conversion (Demirbas, 2006 and 2007). Generally, moisture and free fatty acids content of the feedstock must be reduced or lowered (<0.5% for homogenous process) to avoid its undesirable effect on the catalyst and the transesterification reaction (Ma & Hanna, 1999; Sharma *et al.*, 2008). The higher the acidity of the oil, smaller is the conversion efficiency. Both, excess as well as insufficient amount of catalyst may cause soap formation (Meher *et al.*, 2004). In homogenous transesterification process especially, moisture and free fatty acids in the feedstock could lead to a side saponification reaction, which produces soap and eventually emulsion (Bkansedo, 2009). The resulting soap and emulsion can induce an increase in viscosity, formation of gels and foams, and make the separation and purification process of the final product, which eventually leads to loss of triglyceride and product (biodiesel)(Ghadge and Raheman; 2005). The presence of water has a greater negative effect on transesterification than that of the FFAs. Ma *et al.* (1998) investigated the transesterification of beef tallow catalyzed by sodium hydroxide (NaOH) in presence of FFAs and water. These authors reported that water and free fatty acid contents of beef tallow had to be maintained below 0.06 wt% and 0.5 wt%, respectively. Using NaOH catalyzed transesterification, methyl esters can generally be prepared in high yields for low FFA oils, being nearly quantitative for the palm oils containing <1% FFA. For example, the yield of methyl esters from RBD (refined, bleached and deodorized) palm oil with about 0.05% FFA was 98% (May,2004).

2.4.2. Effect of Alcohol/oil Molar Ratio and Alcohol Type

One of the most important factors that affect the yield of ester is the molar ratio of alcohol to triglyceride. Although the stoichiometric molar ratio of methanol to triglyceride for transesterification is 3:1, higher molar ratios are used to enhance the solubility and to increase the contact between the triglyceride and alcohol molecules (Ma & Hanna, 1999; Noureddini *et al.*, 1998). Higher molar ratios result in greater ester conversion in a shorter time. Hence, in the transesterification process more alcohol is preferred to shift the equilibrium to form higher yield of fatty acid alkyl esters. Molar ratio of alcohol to oil at 6:1 is considered as the standard ratio (Fukuda *et al.*, 2001; Gerpen, 2005). However, other researches also shown that molar ratio of alcohol to oil from 5:1 (Alamu *et al.*, 2008) up to 8:1 (Ramadhas *et al.*, 2005), 9:1 (Sahoo *et al.*, 2007), 12:1 (Meher *et al.*, 2006), and higher could also be used as the optimum ratio for oil to methanol, depending on the quality of feedstock and method of the transesterification process.

According to Meher *et al.* (2006) the reaction was faster with higher molar ratio of methanol to oil whereas longer time was required for lower molar ratio (6:1) to get the same conversion. In their research, the molar ratio of methanol to oil, i.e., 6:1, 9:1, 12:1, and 24:1, were examining for optimizing biodiesel production from Karanja oil. However, when the ratio of oil to alcohol is too high, it could give adverse effect on the yield of fatty acid alkyl esters. Some researchers reported that addition of large quantity of methanol, i.e. at ratio of 1:70 and 1:84 could slow down the separation of esters and glycerol phases during the transesterification process, therefore affecting the final yield of fatty acid alkyl esters (Miao & Wu, 2006).

2.4.3. Type and Amount of Catalyst

The type and amount of catalyst required in the transesterification process usually depend on the quality of the feedstock and method applied for the transesterification process. For a purified feedstock, any type of catalyst could be used for the transesterification process (Gerpen, 2005). Catalysts used for the transesterification of triglycerides are classified as alkali, acid,enzyme or hete rogeneous catalysts, among which alkali catalysts like sodium hydroxide, sodium methoxide, potass ium hydroxide and potassium methoxide are more effective, much faster than acid-catalyzed transesterification and is most often used commercially (Ma & Hanna, 1999). May (2004) studied the effect of different catalysts types on methanolysis of palm oil with a low FFA content of <0.1%. In that study, it was concluded that NaOH and KOH are effective catalysts. Stavarache *et al.* (2005) investigated the effect of different catalyst concentrations on base-catalyzed transesterification during bio-diesel production from vegetable oil by means of ultrasonic energy. The best yields were obtained when the catalyst was used in small concentration, i.e., 0.5% wt/wt of oil. Meneghetti *et al.* (2006) investigated the effect of different catalyst types at different temperatures during the production of free and bound ethyl ester (FAEE) from castor oil. Results from that study showed that hydrochloric acid is much more effective than sodium hydroxide at higher reaction temperatures.

The yield of fatty acid alkyl esters generally increases with increasing amount of catalyst (Demirbas, 2007; Fukuda *et al.*, 2001; Ma & Hanna, 1999). This is due to availability of more active sites by additions of larger amount of catalyst in the transesterification process. However, on economic perspective, larger amount of catalyst may not be profitable due to cost of the catalyst itself.

2.4.4. Reaction Time and Temperature

Freedman et al (1986) observed the increase in fatty acid esters conversion when there is an increase in reaction time. The reaction is slow at the beginning due to mixing and dispersion of alcohol and oil. After that the reaction proceeds very fast. However the maximum ester conversion was achieved within < 90 min. Further increase in reaction time does not increase the yield product i.e. biodiesel/mono alkyl ester (Leung and Guo, 2006; Alamu *et al.*, 2007). Ma *et al.* (1998) evaluated the effect of reaction time on transesterification of beef tallow with methanol. Due to the difficulty of mixing and dispersion of methanol into beef tallow, the reaction was very slow during the first minute. From 1 to 5 min, the reaction proceeds very fast. At about 15 min, the production of beef tallow methyl esters reached the maximum value. Current researches have shown that the reaction time for a non-catalytic transesterification process using supercritical alcohol is shorter compared to conventional catalytic transesterification process (Demirbas, 2003 and 2005).

Reaction temperature is another important factor that will affect the yield of biodiesel.Transesterification can occur at different temperatures, depending on the oil used. For the transesterification of refined oil with methanol (6:1) and 1% NaOH, the reaction was studied with three different temperatures (Freedman *et al.*, 1984). After 0.1 h, ester yields were 94, 87 and 64% for 60, 45 and 32 °C, respectively. After 1 h, ester formation was identical for 60 and 45 °C runs and only slightly   lower for the 32 °C run. Temperature clearly influenced the reaction rate and yield of esters (Ma & Hanna, 1999). The boiling point of methanol is 337.8 K. Reaction temperatures higher than this will burn the alcohol and will cause reduced yield. Leung & Guo (2006) indicated that reaction temperature higher than 323 K had a negative impact on the product for neat oil.

2.5. Parameters which define the fuel quality of biodiesel

The characteristics of biodiesel are similar to those of diesel fuel, and, therefore, biodiesel becomes a strong candidate to replace the diesel fuel if the need arises. The fuel properties of biodiesel are influenced at large by the amounts of each fatty acid composition and the alcohol moieties in the feedstock used to produce the esters among which the largest fractions of fatty acids for each of the biodiesel is a potential indication of the rest of the properties (Kinast, 2003; Van Gerpen *et al.*, 2004). The important physical and chemical properties of oil and biodiesel which determined by standard methods were used as follow:

2.5.1. Density (15 °C)

Density is specified in several standards and the purpose is to exclude unrelated materials from being used as biodiesel feedstock (Knothe, 2006). It is also used in the determination of the viscosity of biodiesel. This property is important mainly in airless combustion systems because it influences the efficiency of atomization of the fuel (Felizardo *et al.*, 2006). Density, relative density or specific gravity is a factor governing the quality and pricing of biodiesel. However, this is uncertain indication of its quality unless correlated with other properties. Density was measured with a hydrometer in accordance with ASTM D1298, Standard Test Method for Density, Relative Density (Specific Gravity) (ASTM Standard D1298, 2005).

Biodiesel is generally denser than diesel fuel with sample values ranging from 877 kg/m^3 (tallow methyl ester) to 884 kg/m^3 (soy methyl ester) compared with diesel at 835 kg/m (Sharp, 2000).Thus; density of the final product depends mostly on the feedstock used. The transesterification process has been found in some cases to reduce fuel density (Zheng and Hanna, 1995). Lower density contaminants such as methanol and ethanol also decrease overall density of the fuel.

2.5.2. Viscosity (40 °C)

Viscosity is a measure of the internal fluid friction or resistance of oil to flow, which tends to oppose any dynamic change in the fluid motion (Sivaramakrishnan and Ravikumar, 2012) and is determined by measuring the amount of time taken for a given measure of oil to pass through an orifice of a specified size (Allen *et al*, 1999). It is one of the parameters specified in biodiesel and petro-diesel standards that require compliance since it affects the operation of fuel injection equipment, particularly at low temperatures when the increase in viscosity affects the fluidity of the fuel (Canoira *et al*., 2009; Knothe, 2005). Biodiesel has higher viscosity than conventional diesel fuel. High viscosity leads to poorer atomization of the fuel spray and less accurate operation of the fuel injectors (Strong *et al*., 2004). Knothe and Steidley (2005), reported that the presence of an OH group in ricinoleic acid increases viscosity significantly oil due to hydrogen bonding. Therefore the major reason for reducing viscosity of processing plant oils is to make them suitable for use as biodiesel. Methods of reducing viscosity in addition to transesterification include dilution, microemulsion, pyrolysis and catalytic cracking (Canakci and Gerpen, 2001; Demirbas, 2006; Gemma *et al*., 2004; Hirata and Berchmans, 2007).

Ranges of acceptable kinetic viscosity at 40 °C are 1.9 – 6.0 mm2/s as required by ASTM D6751 specification. The kinetic viscosity of fatty compounds (such as those found in biodiesel fuel) is significantly influenced by compound structure, including chain length, the position, number, and nature of double bonds, and the nature of oxygenated moieties (Knothe *et al*., 2005). Biodiesel viscosity is also dependent on temperature. It is reported that biodiesel viscosity can be calculated in the range from 273 K to 303 K with one equation (Kerschbaum and Rinke, 2004).

2.5.3. Gross calorific value

Calorific value of a fuel is the thermal energy released per unit quantity of fuel when the fuel is burned completely and the products of combustion are cooled back to the initial temperature of the combustible mixture. It measures the energy content in a fuel. This is an important property of the bio-diesel that determines the suitability of the material as alternative to diesel fuels (Sivaramakrishnan and Ravikumar, 2012). The heat content of vegetable oils and their alkyl esters is nearly 90% that of DF and the heats of combustion of fatty esters and triacylglycerols are in the range of ~1300– 3500 kg cal/mol for C8– C22 fatty acids and esters (Freedman and Bagby, 1989). One of the most important determinants of heating value is moisture content. The calorific value of vegetable oils and their methyl esters were measured in a bomb calorimeter according to

ASTMD240 standard method (Sivaramakrishnan and Ravikumar, 2012; Parr, 1987)). Benzoic acid was used to standardize the calorimeter. One gram of sample was taken in a crucible and made into a pellet and the initial weight was noted. It was placed in the bomb, which is pressurized to 18 atm of oxygen (Mesfin, 2008). For purposes of comparison, the literature value for the heat of combustion of cetane is 2559.1 kg-cal/mol (at 20 °C); thus it is the same range as fatty compounds (Weast *et al*, 1985–1986).

2.5.4. Cloud point (CP)

Low temperature operability of biodiesel fuel is an important aspect from the engine performance standpoint in cold weather conditions (ASTM Standard D2500, 2005). There are several tests that are commonly used to determine the low temperature operability of biodiesel. CP is one of these tests and is included as a standard in ASTM D6751. It is the temperature at which crystals of organic matter in the biodiesel are visualized when measured by lowering its temperature according to the procedures described by ASTM D2500. ASTM D6751 requires the producer to report the cloud point of the biodiesel sold, but it does not set a range as the desired cloud point is determined by the intended use of the fuel (ASTM Standard D2500, 2005).

While operating an engine at temperatures below oil's cloud point, heating will be necessary in order to avoid waxing of the fuel. Small crystals of fuel begin to form in the liquid causing haziness as the sample is cooled. It is an indicator of the utility of petroleum oil for some applications. In the case of biodiesel, the haze is made up of crystallized fuel molecules, specifically crystallized stearic and/or palmitic methyl esters (Knothe and Dunn, 2001).

2.5.5. Cetane Number

Cetane Number (CN) or aniline point is a relative measure (the scale) of the interval between the beginning of injection and auto-ignition of the fuel (conceptually similar to the octane scale used for gasoline). The CN is the primary specification measurement used to match fuels and engines (Van Gerpen *et al.*, 2004). In a compression ignition diesel engine the cetane number is the measure of ignition promotion. In a spark ignited gasoline engine the ignition quality of gasoline is measured by the octane number which is a rating of ignition delay (Midwest Biofuels, 1994). The cetane number is not to be confused with the cetane index, which is not applicable to biodiesel. Cetane indices predict the cetane number from equations derived for petroleum distillates only. The cetane number of biodiesel depends on the distribution of fatty acids in the original oil or fat from which it was produced. The longer the fatty acid carbon chains and the more saturated the molecules, the higher the cetane number (Van Gerpen *et al.*, 1996).

The cetane number of a fuel reflects its ignition delay. A fuel of higher cetane number gives lower delay period and provides smoother engine operation (Mittelbach and Gangl, 2001). Therefore, high cetane number is desirable for engine fuel. Biodiesel has a higher CN than petrodiesel because of its higher oxygen content (Mittelbach and Gangl, 2001). Cetane numbers of biodiesels differ depending on the respective oil sources. Cetane numbers of biodiesel from soybean oil methyl esters lie between 45.8 and 56.9, and that from rapeseed oil methyl esters lie between 48 and 61.8. Cetane increases with chain length, decreases with the number of double bonds, and decreases as double bonds and carbonyl groups move towards the centre of the chain. Increasing cetane number of biodiesel has been shown to reduce nitrogen oxides emissions (Van Gerpen *et al.*, 2004).

2.5.6. Iodine Value (IV)

Iodine value (number) is a measure of the total unsaturation within a mixture of fatty acids, and is expressed in grams of iodine which react with 100 grams of biodiesel. Engine manufacturers have argued that fuels with higher iodine number tend to polymerize and form deposits on injector nozzles, piston rings and piston ring grooves when heated (Kosmehl and Heinrich, 1997). Moreover, unsaturated esters introduced into the engine oil are suspected of forming high-molecular compounds which negatively affect the lubricating quality, resulting in engine damage (Schaefer *et al.*, 1997). Biodiesel viscosity is directly correlated to the iodine number of biodiesel for biodiesel with iodine numbers of between 107 and 150 (Prankl and Worgetter, 1996).

2.5.7. Flash point

The flash point is a measure of the lowest temperature at which application of the flame causes the vapor above the sample to ignite, i.e., it is a measure of the tendency of a sample to form a flammable mixture with air (Van Gerpen *et al.*, 2004).It is the lowest temperature at which fuel emits enough vapors to ignite (ASTM Standard D93, 2008). FP varies inversely with the fuel's volatility (Sivaramakrishnan and Ravikumar, 2012). It specifies the temperature to which a fuel needs to be heated for the vapour and air above the fuel could be ignited (Sarma, 2005). Minimum flash point temperatures are required for proper safety and handling of diesel fuel. Fire point is the lowest temperature at which a specimen will sustain burning for 5 seconds (Sivaramakrishnan and Ravikumar, 2012). Biodiesel has a high flash point; usually more than 150°C, while conventional diesel fuel has a flash point of 55-66°C (Knothe *et al.*, 2005). Furthermore, the flash point of methyl ester fuels is higher than that of ethyl esters. If methanol, with its flash point of 12°C is present in the biodiesel the flash point can be lowered considerably (Mallinckrodt Baker Inc., 2009). Flash points of the samples were measured in the temperature range of 60 to 190°C by an automated Pensky-Martens closed cup apparatus (Sivaramakrishnan and Ravikumar, 2012).

2.5.8. Water and sediment

Water and sediment is a test that determines the volume of free water and sediment in middle distillate fuels having viscosities at 40 °C in the range 1.0 to 4.1 mm^2/s and densities in the range of 700 to 900 kg/m3 (Van Gerpen *et al.*, 2004).

This test is a measure of cleanliness of the fuel. For B100 it is particularly important because water can react with the esters, making free fatty acids, and can support microbial growth in storage tanks. Water is usually kept out of the production process by removing it from the feedstocks. However, some water may be formed during the process by the reaction of the sodium or potassium hydroxide catalyst with alcohol. If free fatty acids are present, water will be formed when they react to either biodiesel or soap. Finally, water is deliberately added during the washing process to remove contaminants from the biodiesel. This washing process should be followed by a drying process to ensure the final product will meet ASTM D 2709 (Van Gerpen *et al.*, 2004).Sediments may plug fuel filters and may contribute to the formation of deposits on fuel injectors and other engine damage. Sediment levels in biodiesel may increase over time as the fuel degrades during extended storage (Van Gerpen *et al.*, 2004).

2.5.9. Carbon Residue

An important indicator of the quality of biodiesel is the carbon residue, which corresponds strictly to the content of glycerides, free fatty acids, soaps, remaining catalyst and other impurities (Mittelbach, 1996).This parameter indicates the tendency of the fuel to form carbon deposits in an engine. Carbon residue which is formed by decomposition and subsequent pyrolysis of the fuel components can clog the fuel injectors. ASTM D6751 includes carbon residue as a standard for biodiesel. The maximum allowable carbon residue for biodiesel is 0.050 % by mass (ASTM Standard D6751, 2009).

"In petroleum products, the part remaining after a sample has been subjected to thermal decomposition..." is the carbon residue. The carbon residue is a measure of how much residual carbon remains after combustion. The test basically involves heating the fuel to a high temperature in the absence of oxygen. Most of the fuel will vaporize and be driven off, but a portion may decompose and pyrolyze to hard carbonaceous deposits. This is particularly important in diesel engines because of the possibility of carbon residues clogging the fuel injectors (Van Gerpen *et al.*, 2004).

2.5.10. Sulfated ash

Sulfated ash is the residue remaining after a fuel sample has been carbonized, and the residue subsequently treated with sulfuric acid and heated to a constant weight. This test monitors the mineral ash residual when a fuel is burned (Van Gerpen *et al.*, 2004). Ash is formed from abrasive solids, soluble metallic soaps and unremoved catalysts remaining in the biodiesel. Combustion in the engine oxidizes these materials to ash. The ash content is specified in standards either as ash (oxidized) content or sulfated ash content. There is a correlation between the sulphated ash content and the phosphorous content of the oil (Mittelbach, 1996). For biodiesel, this test is an important indicator of the quantity of residual metals in the fuel that came from the catalyst used in the esterification process. Producers that use a base catalyzed process may wish to run this test regularly. Many of these spent sodium or potassium salts have low melting temperatures and may cause engine damage in combustion chambers (Van Gerpen *et al.*, 2004).

2.5.11. Acid value

Acid value or neutralization number is a measure of free fatty acids contained in a fresh fuel sample and of free fatty acids and acids from degradation in aged samples. If mineral acids are used in the production process, their presence as acids in the finished fuels is also measured with the acid number. It is expressed in mg KOH required to neutralize 1g of biodiesel. It is influenced on the one hand by the type of feedstock used for fuel production and its degree of refinement. Acidity can on the other hand be generated during the production process. The parameter characterizes the degree of fuel ageing during storage, as it increases gradually due to degradation of biodiesel. High fuel acidity has been discussed in the context of corrosion and the formation of deposits within the engine which is why it is limited in the biodiesel specifications of the three regions. It has been shown that free fatty acids as weak carboxylic acids pose far lower risks than strong mineral acids (Cvengros, 1998). Acidity can also result from improper manufacturing, through remaining catalyst (if manufactured in acidic conditions) or excessive neutralization. This is different for diesel, which does not contain these materials (Scherpenzeel, 2000).

2.6. Coffee production and Waste coffee residues

2.6.1. Coffee production in Ethiopia

The word "coffee" comes from the name of a region of Ethiopia where coffee was first discovered – 'Kaffa'. Botanically, coffee belongs to the family *Rubiaceae* in the genus *Coffea*. Although the genus *Coffea* includes four major subsections, 66% of the world production mostly comes from *Coffea arabica* L. and 34% from *Coffea canophora* Pierre ex Froehner (robusta type), respectively. Ethiopia is the home and cradle of biodiversity of Arabica coffee seeds. More genetically diverse strains of *C. arabica* exist in Ethiopia than anywhere else in the world, which has lead botanists and scientists to agree that Ethiopia is the centre for origin, diversification and dissemination of the coffee plant (Mekuria *et al.*, 2001).

In Ethiopia, coffee contributes over 5% of the gross domestic product, 12% of agricultural output, 70% of the foreign exchange earnings and 10% of the government revenues. Besides, 25% of the population is employed in coffee production, processing and marketing (Alemayehu *et al.*, 2007). The potential for coffee production in Ethiopia is very high considering the country's suitable altitude, rainfall, temperatures and fertile soil. It is estimated that the total area covered by coffee is approximately 600,000 hectares (Alemayehu *et al.*, 2007), with a total production of 9.804 million 60-kg bags (African Commodity Report, 2012) and half of the total consumed domestically which can generate 235,296 tons of spent coffee grounds. Ethiopia ranks first in coffee consumption in Africa and the annual national average per capita coffee consumption is 3kg: Small-scale coffee farms contribute about 90% while large scale modern plantations account for the remaining 10% coffee production of the country (Alemayehu *et al.*, 2007).

2.6.2. Waste Coffee Residues (WCRs)

Coffee is one of the most consumed beverages of the world (Mussatto *et al.*, 2011); large quantities of waste generated are a result of the high consumption of coffee worldwide. Spent coffee grounds, i.e. the solid residue remaining after the production of the coffee beverage, are generated in large amounts worldwide, as attested by the over 100 million bags of coffee produced in the world each year (ICO, 2009).They are also the main coffee industry residues with a worldwide annual generation of 6 million tons (Tokimoto *et al.*, 2005). Although the toxic character and presence of organic matter in WCR, the discharge of this residue to the environment and sanitary landfill are disposal forms still performed nowadays, but that avoided. A biological treatment of WCR material with fungal strains from the genus *Penicillium, Neurospora,* and *Mucor* could be an interesting alternative to be performed previous the material elimination to the environment since these fungi are able to release phenolic compounds from the WCR structure, decreasing their toxicity (Mussattoet *et al.*, 2011).In some cases, WCR is used as fuel in industrial boilers of the same industry due to its high calorific power, approximately 5,000 kcal/kg, which is comparable with other agro-industrial residues (Silva *et al.* 1998).

In general it is considered that the oil content in WCR contain 10-20 %*w/w*. Assuming 14 wt. % oil content in WCR, utilizing these oils for biodiesel and production could add approximately 1123 million tons of biodiesel (based on the year 2010) to the world's fuel supply (Al-Hamamre *et al.*, 2012), Thus coffee oil, extracted from coffee grounds was found to be a high quality and cost-effective feedstock for biodiesel production compared to other waste sources. It is less expensive, has higher stability (due to its high antioxidant content), and has a pleasant smell (Kulkarni, 2006).

The remaining solid waste can be used for bioethanol production (Kondamudi *et al.*, 2008; Sendzikiene et al. 2004), compost and also as fuel pellets (Kondamudi *et al.*, 2008).

2.6.2.1. Chemical composition of WCR

As other agricultural and food wastes, WCRs have high volume but low value, which results in reduced global economic interest. They are disposed to landfill, causing major environmental issues. Several studies have been carried out on the last decade dealing with possible strategies to reduce their toxicity and enhance their commercial value, including in production of animal feed, mushroom production, biodiesel, fuel pellets or activated carbons (Mussattoet *et al.*, 2011 and Franca *et al.*, 2009).The main substance present in waste coffee grounds are carbohydrates. As Couto *et al.* (2009) had reported in WCR, hemicellulose is the main compound, 36.7 wt% dry bases. Thus, it indicates that WCR can be used as a source of carbohydrates that can be hydrolyzed to produce reduced sugars, which could be used as carbon sources for several microorganisms. WCR contain on average up to 20 wt% of lipids, 87-93% of them are triglycerides, and 6.5 to 12.5 wt% is diterpene alcohol esters (Calixto *et al.*, 2011). Triglycerides in the WCR can be converted to a similar amount of biodiesel using the transesterification methods (Kondamudi *et al.*, 2008). According to Caetano *et al.* (2012) the composition of the WCR is shown in Table 2.2.

Table 2.2: Composition of waste coffee residue

Parameter	Waste coffee residue
Moisture (%)	12.2
Total carbon (%)	52.2
Total nitrogen (%)	2.1
Protein (gprotein/100 g)	13.3
Ash (%)	1.43
Cellulose (%)	13.8
Soluble lignin (%)	31.9
Total lignin (%)	33.6
Higher heating value (MJ/kg)	4619.2

Source: (Caetano *et al.*, 2012*)*

2.7.Advantages and Disadvantages of Biodiesel

Biodiesel has attracted considerable interest as an alternative fuel or extender for petrodiesel for combustion in compression–ignition (diesel) engines. Biodiesel is miscible with petrodiesel in any proportion and possesses several technical advantages over ultra-low sulfur diesel fuel (ULSD, <15 ppm S), such as inherent lubricity (Alberta., 2006), low toxicity (Kingwell and Plunkett, 2006), derivation from a renewable and domestic feedstock (National Biodiesel Board, 2010), superior flash point and biodegradability, negligible sulfur content (Demirbas, 2005), and lower overall exhaust emissions (Alberta, 2006). Disadvantages of biodiesel include high feedstock cost, inferior storage and oxidative stability, lower volumetric energy content, inferior low-temperature operability versus petrodiesel, and in some cases, higher NOx exhaust emissions (Knothe, 2008). Many of these deficiencies can be mitigated through cold flow improver (Hancsok *et al.* 2008; Moser and Erhan 2008) and antioxidant (Kondamudi *et al.*, 2008; Loh *et al.*, 2006) additives, blending with petrodiesel (Benjumea *et al.*, 2008), and/or reducing storage time (Bondioli *et al.*, 2003).

3. MATERIALS AND METHODS

3.1. Materials and chemicals

The WCR sample which is the residue obtained from brewing coffee was supplied from TOMOCA PLC coffee shop (Addis Ababa, Ethiopia). The waste coffee residue sample was dried in an oven at 105^0C for 24 h to reduce the moisture content. All solvents and analytical grade chemicals were obtained from chemistry department (Addis Ababa University, Ethiopia). Experimental and laboratory work was undertaken in Addis Ababa University; Chemistry Department, Ethiopian Petroleum Supply Enterprise (EPSE), Ethiopian Health and Nutrition Research Institute, Geological Survey of Ethiopia Central Laboratory and Ministry of mines and energy of Ethiopia.

3.2. Experimental

3.2.1. Waste coffee residue (WCR) moisture content determination

The moisture content of the WCR was determined using an oven. The sample was weighed to the nearest 10 g in Petri dishes and then dried at $105 \ ^0C$ for 24 h. It was then cooled in desiccator over silica gel (0% relative humidity) and reweighted. The moisture content was determined as:

$$M = \frac{(W_1 - W_2)}{W_1} \ x \ 100 \ \% \tag{1}$$

Where M, W 1, W 2 are moisture content, wet mass and dry mass, respectively (Yong and Salimon, 2006).

3.2.2. Waste coffee residue (WCR) oil extraction

500g of the dried WCR sample was placed in a 4L round bottom flask of the soxhlet apparatus (fig. 4) and percolated for 6 h using n-hexane as a solvent. The solvent was then removed from the oil using a rotavapor (Buchi RE111 Rotavapor, BUCHI, Flawil, Switzerland) (fig. 5). The solvent was reused for subsequent batch of extracting.

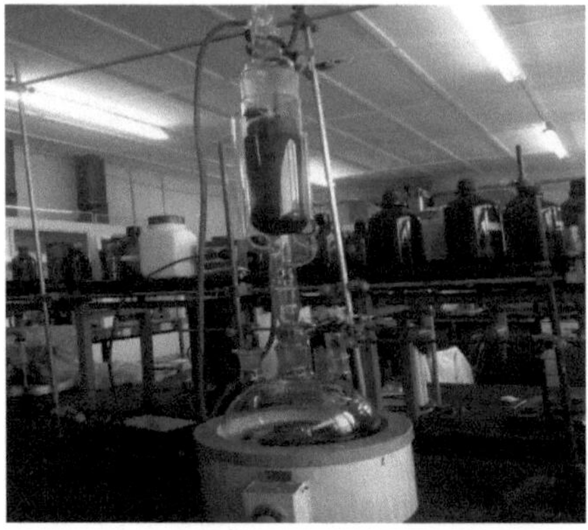

Fig. 4: experimental set up for Soxhlet Extraction (Photo: Author)

Fig. 5: Rotavapor used to separate hexane from the extracted WCR oil (Photo: Author)

The weight of the oil was recorded after removing the hexane from the crude extract. The oil yield was determined using equation (2).

$$\text{Waste coffee residue oil content} = \frac{W_o}{Ws} \; x \; 100 \; \% \tag{2}$$

Where: W_o = weight of oil extracted

Ws = weight of the sample Weight of sample (dry base)

Fig. 6: WCRs before extraction (left) and after extraction (right) (Photo: Author)

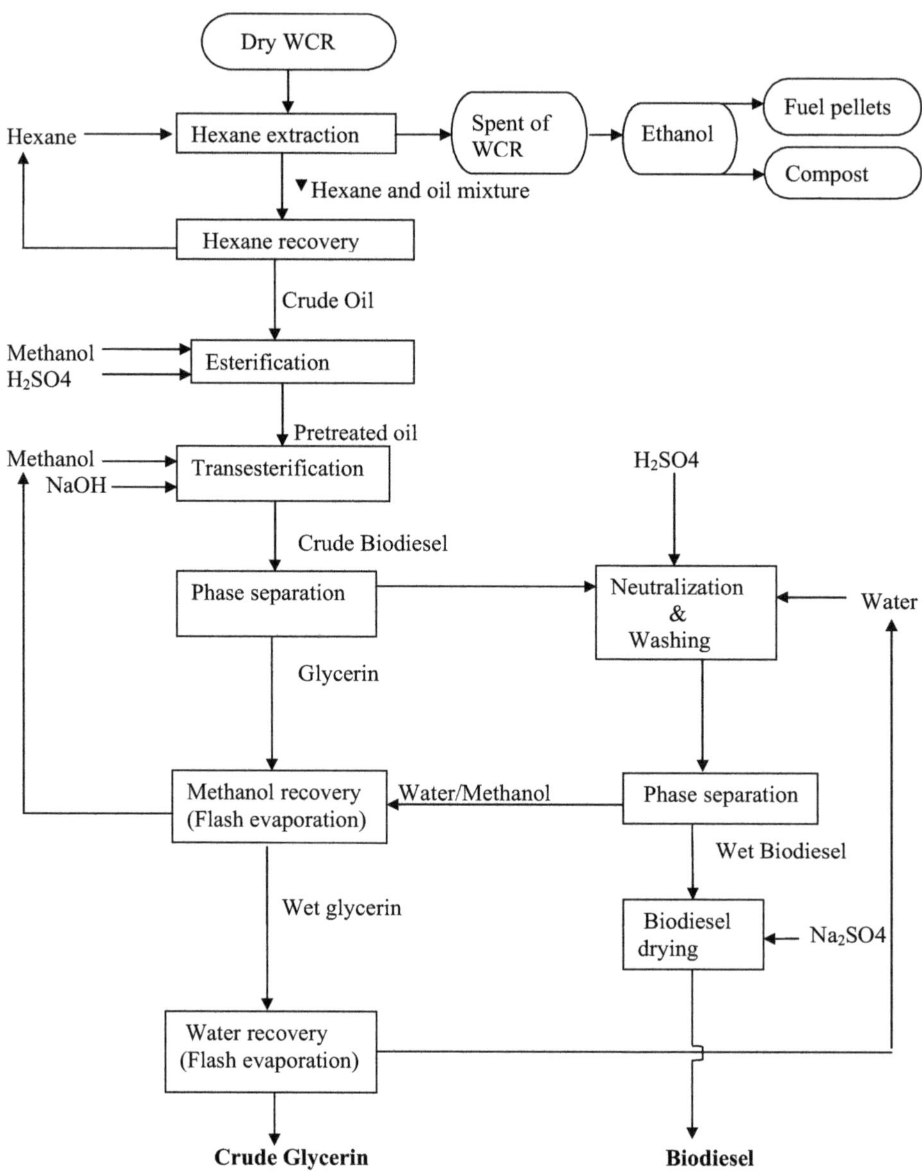

Fig. 7: General Flow chart of the study (prepared flow chart)

3.2.3. Physicochemical parameters of WCRs oil

The WCR oil was characterized for its physical and chemical properties. Standard methods applied for WCR oil are presented in Table 3.2. The details on equipment and procedures for peroxide value and saponification value are presented as follows and for other parameters the same as in the ester characterization part (3.2.7);

3.2.3.1. Determination of Saponification Value

2 g of the oil was poured in to conical flask and exactly 25 mL of 0.5 N Ethanol potassium hydroxide solutions was added to it, a reflux condenser was attached and the resulting mixture was heated in boiling water for 60 minutes. The excess potassium hydroxide was titrated with 0.5 N hydrochloric acid using phenolphthalein as indicator while still hot. The resulting end point was obtained when the pink colour changed into colorless. A blank determination was carried out under the same condition and the S.V calculated as below:

$$\text{Saponification value (SV)} = \frac{(A - B) \times 28.05}{\text{Weight of sample}} \qquad (3)$$

Where;   B – ml of HCl required by blank
S –ml of HCl required by sample
28.05= constant

3.2.3.2. Determination of Peroxide value

5 g of sample was weighed into clean dried 250 mL conical flask. 30 mL of the solvent mixture (i.e, glacial acetic acid and chloroform in the ratio 18:12) was added to it and the air above the liquid was displaced with carbon dioxide. Then 1 mL of the Potassium iodide solution was added flask to get homogeneous solution. The solution was allowed to stand in the dark for 1 minute and then 30 mL water was added. Resulted mixture was titrated with 0.02 M Sodium thiosulphate using freshly prepared 1% starch solution, a blank titration was carried out at the sample time, and the peroxide value was calculated using the relationship below (Eq. 4).

$$\text{Peroxide value} \ = \ \frac{T \times M \times 1000}{W} \qquad\qquad (4)$$

Where: T = titre value of Na2S$_2$O$_3$ = sample titre – blank titre

M= molarity of Na$_2$S2O$_3$

W= weight of sample

Table 3.1: Standard test methods for physicochemical properties of WCR oil

Property	Units	Test method
Density (15 °C)	g/cm^3	ASTM D1298
Kinematic Viscosity(40 °C)	mm²/s	ASTM D 445
Gross calorific value	MJ/kg	ASTM D 240
Cloud point	°C	ASTM D 97
Iodine number	gI2/100g	EN14111
Water and sediment	%volume	ASTM D2709
Acid value	Mg	ASTM D974
Flash point	°C	ASTM D93
Saponification value	mgKOH/g	AOCS Cd 3-25
Peroxide vale	Meq/kg	D3703

3.2.4. Two-Step Biodiesel Production Process

3.2.4.1. Acid-catalyzed esterification

The WCR oil was heated to 60 °C to homogenize the oil. The reaction was conducted in a 1000 mL two necked round-bottomed flask attached with a reflux condenser and thermometer placed in an oil bath. Oil is mixed with methanol (a molar ratio of alcohol to free fatty acids of 20:1) and significant quantities of H$_2$SO$_4$ (10 wt% of total fatty acids content) (See Appendix 9.1 for calculations). The reactor was stirred at about 600 rpm, at temperature of 60 °C for 2 h (Santori *et al.*, 2012). Then the reaction product mixture was poured into a separating funnel and allowed to settle for 24 h. The top layer which is comprised unreacted methanol and water was removed. The remaining FFA was determined before alkali transesterification. The product was then used for the alkaline transesterification. The acid pre-treatment loss was calculated by (Eq.5).

$$\text{Acid pretreatment loss} \ = \ \frac{\text{weight of pretreated oil}}{\text{Weight of crude oil}} \ x \ 100 \ \% \tag{5}$$

3.2.4.2. Base-catalyzed transesterification

In transesterification process, WCR oil was heated to 100 °C to remove the traces of water present due to esterification. The transesterification reaction of WCR oil was carried out in a 1000 mL round-bottomed glass flask, with anhydrous methanol in molar ratio methanol to oil 6:1, using sodium hydroxide (NaOH) as catalyst in amount 1% (w/w) (See appendix 9.2 for calculations). The reaction was taken place at 60 °C for 2 h. By the end of the experiment the reaction product was then poured into a separating funnel for glycerol and methyl ester separation, allowed to settle for 24 h (Deligiannis *et al.*, 2011). After the two phases separation, the excess alcohol in each phase was removed by flash evaporation at 90 °C placed on oil bath. The methanol was recovered and re-used.

Fig.8: Transesterification of WCR oil (Photo: Author)

3.2.4.3. Purification

The Biodiesel produced was quite caustic with a pH of between 8.0 and 9.0. Washing with water neutralizes the product. Therefore, the methyl ester was purified by washing gently with warm (55 °C) deionized water to remove residual catalyst, glycerol, methanol and soap using a centrifuge. A small amount of sulphuric acid (H_2SO_4) was used in the second washing to neutralize remaining soaps and other catalyst impurities. The washed methyl ester was then dried over the heated anhydrous sodium sulphate (Na$_2$SO$_4$). Solid traces from the methyl ester were removed through a filtration process. The dried waste coffee residue methyl ester was then bottled and kept for characterization studies.

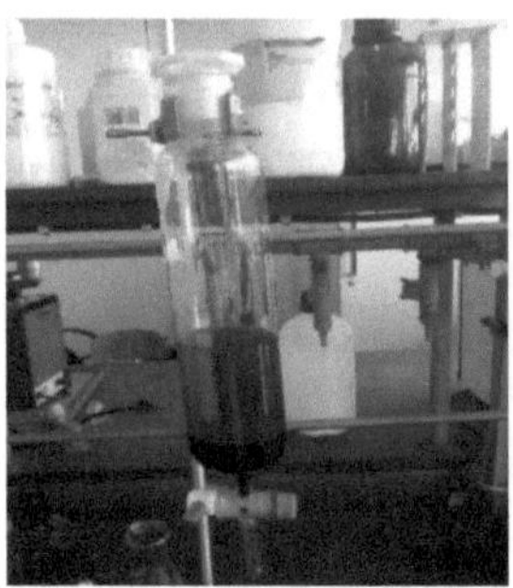

Fig.9: During phase separation **Purification stage** **Purified WCR biodiesel** (Photo: Author)

3.2.5. Biodiesel yield

The biodiesel yield (% wt) after the post-treatment stage, relative to the amount of WCR oil poured into the reactor, was calculated from the methyl ester and WCR oil weight.

$$\text{Biodiesel yield} = \frac{\text{Weight of Biodiesel produced}}{\text{Weight of WCRs oil}} \times 100\ \% \tag{6}$$

3.2.6. Characterization of WCR Biodiesel

The biodiesel were characterized for their physical and chemical properties. Standard methods applied for ester characterization are presented in Table 3.3. Their details as follow;

3.2.6.1. Determination of specific gravity/density (ASTM D1298) by hydrometer method

The sample was taken into 250 mL graduated cylinder and its temperature was recorded. Hydrometer was used to measure the SG of the WCR oil and biodiesel. The hydrometer was

50

immersed in to the 250 mL graduated cylinder which is holding the sample and the SG reading was recorded. Then, temperature correction factors were applied to convert the measured specific gravity to the reference temperature of 60/60 °F and density was taken at temperatures of 15 °C and 20 °C. The observed hydrometer reading was converted to density and relative density using the appropriate parts of the petroleum measurement tables in guide to ASTM D 1250-80 (1980). The final value was recorded as density in g/mL at the reference temperature of 15 °C to the nearest 0.0001.

3.2.6.2.Determination of Kinematic Viscosity (40 °C)

The viscosity of the sample was measured at 40 °C using a glass capillary viscometer. The temperature of the viscometer bath was adjusted to 40 °C. The viscometer together with its content (sample) was allowed to remain in the bath still the test temperature to reach (40 °C). The sample was then allowed to freely flow and the time required for the meniscus to pass from the first to the second timing marks was noted with a stop watch. The procedure was repeated to make a second measurement of flow time and the average of these determinations was used to calculate the kinematic viscosity. It was calculated as follow;

$$\mathbf{v = C\ x\ t} \tag{7}$$

Where v = kinematic viscosity, mm^2/s

C = calibration constant of the viscosity, $(mm^2/s)/s$

t = mean flow time, s

Fig.10: Cannon- Fenske glass capillary viscometer tube with samples in the SETA KV-8 viscometer bath (Photo: Author)

3.2.6.3. Determination of Acid value

Acid value was determined by the acid–base titration technique. Potassium hydroxide (KOH) was used as a standard alkali solution. 20 g of the sample was weighed into 250 mL Erlenmeyer flask. 1000 mL of titration solvent (which is a mixture of 495 mL Isopropanol, 500 mL of toluene and 5 mL distilled water per liter of solution) and 0.5 mL of indicator solution (P-Naphtholbenzein) were added to the flask. The mixture was stirred until the sample was entirely dissolved by the solvent. Without delay, the mixture was titrated with the standardized 0.067 M KOH titrant. The end point was noted when the color of the indicator changed to a stable green-black. Blank titration was performed on a 1000 mL of the titration solvent and 0.5 mL of the indicator solution by similar procedure. The Acid value was calculated by the following equation:

$$\text{Acid Value} = \frac{(A - B) \times M \times 56.1}{W} \tag{8}$$

Where: A = milliliters of titrant required for titration of the sample

B = milliliters of titrant required for titration of the blank

M = Molarity of standard KOH and

W = grams of sample used

3.2.6.4. Determination of Gross calorific value

An oxygen adiabatic bomb calorimeter was employed to measure the heating values of the sample. A 1 g of the sample was weighed and kept in the metal combustion capsule. The capsule was then placed on the bomb calorimeter where 10 cm of fuse wire was attached. The bomb calorimeter was sealed by closing the screw cap on top of the head firmly. The bomb calorimeter was pressurized to 30 atmosphere of oxygen. The bomb calorimeter was placed in a 2000 mL bucket containing measured distilled water with its feet spanning the circular boll in the bottom of the bucket. The ignition circuit was connected and the bomb calorimeter cover was closed. After ignition the temperature rise was noted every minute till a constant temperature was reached. The bomb calorimeter was removed from the bucket and the pressure was released. Washed all interior surfaces of the capsule with distilled water and collected the washings in a beaker. Removed all unburned pieces of fuse wire from the bomb calorimeter electrodes, measured their combined length in cm. subtracted their length from the initial length of 10 cm. Titrated the capsule washing with standard 0.0709 N Na_2co_3 solution using methyl as an indicator until red color change in to orange. The calorific value was finally calculated based on the following equation;

$$Hg = \frac{TW - e_1 - e_2 - e_3}{M}$$
(9)

Where, Hg = Gross heat of combustion

T = Temperature difference (°C), (final temperature – initial temperature)

W = Energy equivalent of calorimeter in Cal/°C (2420 Cal/°C)

M = Mass of sample in grams

e_1 = Correction in calories for heat of formation of HNO_3

e_2 = Correction in calories for heat of formation of H_2SO_4

e_3 = Correction in calories for heat of combustion of fuse wire

3.2.6.5.Determination of Cloud point (ASTM D 2500)

40 mL of liquid sample was placed in a test jar. The jar was then closed tightly with a cork carrying a test thermometer (fig.11). The test jar was placed on the disc located on a gasket inside the jacket. The jacket was then filled with layers of ice and sodium sulfate in order to cool down the jacket. At each test thermometer reading that is multiple of 1 °C, remove the test jar from the jacket quickly but without disturbing the specimen, inspected for cloud, and replaced in the jacket (3 seconds). Reported the cloud point, to the nearest 1 °C, at which any cloud was observed at the bottom of the test jar which was confirmed by continued cooling.

Fig.11: The Peltier device apparatus (Photo: Author)

Table 3.2: Test methods to characterize WCR Biodiesel

Property	Units	Test method
Density (15 °C)	g/cm³	D1298
Kinematic viscosity(40 °C)	mm²/s	D445
Gross calorific value	MJ/kg	D 240
Cloud point	°C	D97
iodine number	gI2/100g	EN14111
water and sediment	%volume	D2709
Ash content	W%	D874
acid value	Mg	D974
Carbon residue	W%	D189
flash points	°C	D93
Distillation, 90% recovery	°C	D86
Copper corrosion	Max.	D130
Cetane number	Min.	D613

3.2.6.6.Determination of Water and sediment

Water and sediment test were determined according to ASTM D 2709 Standard Test Method (Table 3.3). The sample container and its contents had equilibrated to room temperature, and the sample was agitated by hand for 10 minutes to ensure homogeneity. 100 mL sample was placed in two identical tubes and centrifuged at 1870 rpm for 10 minutes at room temperature (fig. 12). After 10 minutes, the volume of water and sediment which was settled into the tip of the centrifuge tube was recorded to the nearest 0.005 mL and reported as the volumetric percent water and sediment.

3.2.6.7. Determination of Cetane index/ number

Cetane index is used to determine the ability of fuel to ignite quickly after being injected. This is an important property for biodiesel ester as a diesel fuel substitute. The cetane index was measured as per the correlation given by Krisnangkura (1986) as shown below:

$$\textbf{Cetane index} = \mathbf{46.3} + \frac{\mathbf{5458}}{\textbf{saponification value}} - (\mathbf{0.225 \times Iodine\ value}) \qquad (10)$$

Cetane number is not showed much difference from cetane index. The correlation reported by Patel (1999) is as given below:

$$\textbf{Cetane number} = \textbf{cetane index} - \mathbf{1.5\ to} + \mathbf{2.6} \qquad (11)$$

3.2.6.8. Determination of Iodine Value

0.25 g of the sample was placed in a 250 mL conical flask and 10 mL of chloroform was added and dissolved the sample. 25 mL of hannous solution was added to this flask in a fume chamber. Closed the flask mouth with Stopper and the content of the flask was vigorously swirled. The flask was then placed in the dark for 30 minutes. At the end, 10 mL of 15% aqueous Potassium iodide and 100 mL of water were added. The content was titrated with 0.1 N Sodium thiosulphate solutions. Few drops of 1% starch indicator were added and the titration continued until coloration disappeared after vigorously shaking. A blank determination was carried out under the same conditions. The titre value was recorded and used to calculate the I.V as indicated below;

$$\text{Iodine Value (I.V)} = \frac{1.269\ (V_1 - V_2)}{W} \qquad (12)$$

Where; V_1 = volume of sodium thiosulphate used for blank

V$_2$ = volume of sodium thiosulphate used for determination

W = weight of sample

1.269= Constant

3.2.6.9.Determination of flash point by Pensky-Martens closed cup tester

Flash point was determined Based on ASTM D93 using a Pensky- Martens closed cup tester (fig.13). In this method, the sample was filtered and taken (75 mL) into a brass test cup to the filling mark inside of it which is then fitted with a cover. The appropriate thermometer was also inserted to record the observed flash point. The sample is then heated and stirred, and an ignition source (i.e. a small flame) is directed into the test cup at regular temperature intervals until a flash is detected. The observed flash point was corrected to 760 mmhg pressure (Eq.13).

$$\text{Corrected flash point} = (C + 0.033 (760 - P) \tag{13}$$

Where;   C = observed flash point in °C, and

P = ambient barometric pressure in mmHg.

Fig.13: Pensky- Martens closed cup tester (Photo: Author)

3.2.6.10.   Determination of conradson carbon residue

Conradson carbon residue was determined as per the ASTM D189 (Table 3.2). The crucible was heated to the temperature of 800 °C for a minimum of 10 minutes and cooled to room temperature in a desiccator, and weighed. The sample was mixed thoroughly, filtered and weighed (10 g) to the nearest 5 mg in to a porcelain crucible. The crucible was placed in the center of the Skidmore. The sand in the large sheet iron crucible was labeled and the Skidmore was set on it in the exact center of the iron crucible. The covers was applied to both the Skidmore and the iron crucible, the one on the latter fitting loosely to allow free exit to the vapors as formed. Heat was applied with strong flame, so that the pre- ignition period was 10 ± 1.5 minutes. The period of burning the vapors was 13 ± 1 minutes. When the vapors ceased to burn and no further blue flame was observed, the burner was readjusted and held the heat as at the beginning so as to make the bottom and lower part of the sheet iron crucible a cherry red, and maintained for exactly 7 minutes. The burner was removed and the apparatus was cooled for 15 minutes. The crucible was removed, placed in the desiccator, cooled, and weighed. The carbonaceous type residue remaining was reported as a percentage of the original sample as carbon residue (micro). The mass percentage carbon residue in the original sample was calculated as follows;

$$\%Carbon\ residue = \frac{A \times 100}{W} \tag{14}$$

A = mass of carbon residue (g) and

W = mass of sample used (g)

Fig.14: Carbon residue test apparatus assembly at pre ignition stage (Photo: Author)

3.2.6.11. Determination of Ash content

The crucible was heated to the temperature of 800 °C for a minimum of 10 minutes and cooled to room temperature in a desiccator, and weighed. Then the well mixed 20 g of the sample was weighed into the crucible. The crucible with the sample was heated on a heating plate until the contents were ignited by the flame (fig.15). The temperature of the crucible was maintained for the sample to burn at a uniform and moderate rate until only carbonaceous residues were left in the crucible. The carbonaceous residue was reduced to ash by heating in a muffle furnace at a temperature of 800 °C. The crucible was cooled for 25 minutes at room temperature in a desiccator and its weight was taken. Finally, the ash content of the biodiesel was calculated as follows:

$$\%\text{Carbon residue} = \frac{w}{W} \times 100 \tag{15}$$

Where   w = mass of Ash (g)

W = mass of sample (g)

Fig.15: The samples in crucible at ignition stage (Photo: Author)

3.2.6.12. Determination of Copper strip corrosion

The copper strip corrosion test covers the detection of the corrosiveness of the fuel on copper. This test is based on the effect of the test sample on a polished copper strip. The polished copper strip was immersed in 30 mL of the sample being tested and heated for 3hours at 100 °C under oil bath. At the end of heating period, the copper strip is removed and washed. The color and tarnish level are assessed against the corrosion standard according to ASTM D130 (fig.16).

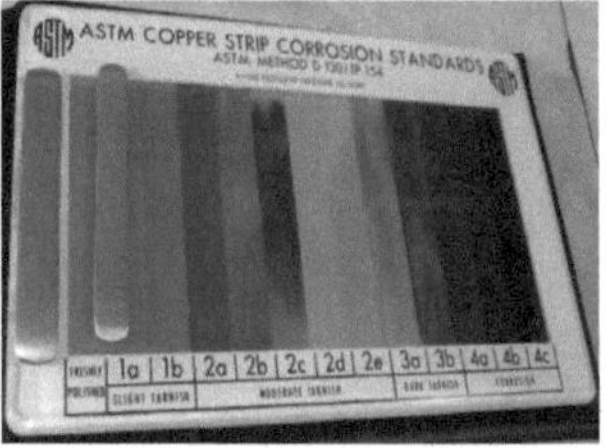

3.2.6.13.   Determination of distillation characteristics

Distillation temperature was determined as per the ASTM D86 (Table 3.3). A 100 mL sample was placed in a distillation flask at ambient pressure. The thermometer bulb was centered in the neck, and the lower end of the capillary was leveled with the highest point on the bottom of the inner wall of the vapor tube of the distillation flask. Then the flask vapor tube was fit, provided a cork tightly in to the condenser tube. The flask was adjusted in a vertical position so that the vapor tube extended in to the condenser tube for a distance from 25 to 50 mm. The flask support metal shield or enclosure was raised and adjusted to fit it snugly against the bottom of the flask. Then the receiving cylinder was placed without drying its contents to its temperature controlled bath under the lower end of the condenser tube to recover the distillate (fig.17). In the interval between the initial boiling point (IBP) and the final boiling point (FBP), the distillation temperature readings at prescribed volume percentages recovered was observed and recorded to the nearest 0.5 °C. The temperature readings were corrected to 760 mmhg pressure by means of the following equation.

$$Cc = \quad 0.00012(760 - P)\,(273 + tc) \tag{16}$$

Where: Cc = corrected temperature reading in, °C,

t_c = the observed temperature reading, °C and

P = barometric pressure, prevailing at the time and location of the test, mmHg.

Fig.17: Setastill distillation apparatus (Photo: Author)

3.2.7. Fatty acid composition of WCR biodiesel

Fatty acid composition of the synthesized alkyl ester was determined by gas chromatography. Gas chromatograph (DANI GC 1000) equipped with flame ionization detector (FID) was employed during fatty acid determination. The GC was calibrated by injecting standards at varying concentrations. 1μL of the sample was injected in to GC, equipped with a capillary column of EC TM-5 (25 m x 0.53 mm x 1μm). The oven starting temperature was 50 °C and kept for 2 minutes. Then 15 minutes hold time with heating rate of 4 °C /minute to 250 °C. Nitrogen at (1 mL/minute) was used as carrier gas at a flow rate of 1.25 bars was adjusted.

3.2.8. Production of Bioethanol from the solid waste remaining after oil extraction of WCR (Spent of WCR)

Spent of WCR was hydrolyzed by refluxing (Fig.18 a), a solid to liquid ratio of 1:10, using dilute sulfuric acid (each of 1, 2 and 3 M) and distilled water for 15 minutes at a temperature of 90 °C. Then the liquid fraction of the hydrolysate sample was cooled, filtered with suction filtration and adjusted to PH 5 by adding 2 N sodium hydroxide (NaOH) and concentrated sulphuric acid (H_2SO_4). Fermentation (Fig.18 b) was then followed using yeast (*S. cereviciae*) for 24 h at temperature of 30 °C (Ayele *et al*, 2011). Finally ethanol was separated from the fermented sample by fractional distillation (Fig.18 c) and its concentration was analyzed by FTIR (spectrum 65 PerkinElmer, UK) (Fig.18 d).

Fig.18: a. Hydrolysis   b. Fermentation   c. fractional distillation   d. FTIR reading (Photo: Author)

$$\text{Gram of ethanol (g)} = \frac{(\text{Conc. of ethanol x amount collected (g))}}{100} \qquad (17)$$

$$\text{Yield of ethanol (\%)} = \frac{\text{Gram of ethanol (g)}}{\text{Sample (g)}} \qquad (18)$$

3.2.9. Determination of the quality of the solid residue after Bioethanol production for compost and solid fuel

The major quality properties of the solid waste remaining after Bioethanol production were analyzed using standard test methods (Table 3.3). Their details on procedures are presented as follows;

3.2.9.1. Calorific value

Calorific value of the solid waste was determined using adiabatic bomb calorimeter (ASTM D240). Details in the ester characterization part (3.2.6)

3.2.9.2. Proximate analysis

The moisture content of the sample was determined by using an oven. 1 g of the sample was taken in a silica crucible and dried in an oven to constant weight at 105 °C for 1 h. After 1 h the sample was taken to desiccator and its moisture content was calculated based on difference between weight of sample before and after an oven dry. Volatile matter content of the sample was determined using a muffle furnace maintained at 750 °C for 3 h. Covered Petri dish was used to hold the sample in the furnace for the determination of volatile matter. The volatile matter was also calculated in the same way as the moisture content by weight difference before and after muffle furnace. Ash content of the waste was determined using a muffle furnace maintained at 750 °C for 6 h. Opened Petri dish was used to hold the sample in the furnace for the determination of ash content. Ash content was determined by taking the difference between weight of samples taken and the residue left in respective experiment. Then fixed carbon was calculated based on the difference as below;

$$\text{Fixed carbon} = 100 - (\text{volatile mater} + \text{moisture} + \text{ash}) \tag{19}$$

Table 3.3-Analysis of solid fuel and compost

property		Units
Gross calorific value		MJkg-1
Proximate Analysis	Fixed Carbon	w%
	Moisture	w%
	Ash	w%
	Volatile matter	w%
Total nitrogen		w%
Total carbon		w%

4. RESULTS AND DISCUSSION

4.1. Oil content of waste coffee residue (WCR)

Waste coffee residue sample was collected from TOMOCA PLC (one of the best coffee shop in Addis Ababa). The sample was dried in an oven and the moisture content was found to be 57.6 %w/w (table 4.1). The oil was extracted from the dried sample with hexane using a soxhlet apparatus. The oil content was found to be 19.73 %w/w on dry weight basis (table 4.1), which is relatively higher than that reported for WCR oil (10-15%) depending on the coffee species by Kondamudi *et al.* (2008) and Deligiannis *et al.* (2011). Higher value (28.3 %w/w) by Caetano *et al.* (2012) for WCR oil extracted with 1:1 mixture of Isopropanol and hexane. The variation in the oil yield could be attributed to differences in variety of coffee, solvent type and cultivation climate.

Table 4.1 - Moisture and oil contents of WCR

Run	WCR Moisture content (% w/w)					WCR oil content (%w/w)			
	W1 (g)	w2 (g)	w1-w2 (g)	WCR Moisture content (% w/w)	Av. Moisture content (%w/w)	Ws (g)	Wo (g)	Oil content (%w/w)	Av. Oil content (%w/w)
1	10	4.290	5.710	57.1o		500	87.66	17.53	
2	10	4.242	5.758	57.58	**57.62**	500	110.74	22.15	
3	10	4.183	5.817	58.17		500	100.58	20.12	
4						500	104.28	20.86	
5						500	106.12	21.22	
6						500	89.51	17.90	**19.73**
7						500	96.89	19.38	
8						500	95.97	19.19	
9						500	102.43	20.49	
10						500	99.66	19.93	
11						500	91.36	18.27	

The oil extracted was black and had the aroma of coffee (fig.19). The oil yield of WCR in this study indicates that it is higher than other oil seeds such as, olive (17%), soybeans (18%), cottonseeds (14%), and corn (3.4%) reported by Rossel (1987).

Fig. 19: WCR oil

4.2. Physicochemical characteristics of the extracted WCR oil

WCR oil that was utilized as a feedstock in the production of biodiesel was characterized to determine its physiochemical properties. The results obtained on the properties of the oil were compared with that of American Standard for Testing Materials (ASTM D 6751). The standard values and the results obtained are summarized in table 4.2.

The acid number, expressed as milligrams of potassium hydroxide per gram of sample, is a measure of acidic substance in the oil. It is used as a guide in the quality control as well as monitoring oil degradation during storage. In this study, the acid value of WCR oil was found to be 14.65 mgKOH/g. It was acidic to be directly converted into biodiesel without any pre-treatment, which may indicate a higher degree of oxidation and occurrence of hydrolysis reactions (Knothe, 2007). It exceeded the ASTM and EN maximum standard specifications of 0.5 mgKOH/g. WCR oil has significant effect on transesterification with methanol using alkaline catalyst, and also interferes with the separation of fatty acid ester and glycerol. This indicates that WCR oil would be better converted to biodiesel using the two-stage process, acid catalyzed esterification followed by base catalyzed transesterification. This value is higher than the values 0.62 mgKOH/g and 0.29 mgKOH/g reported for a similar work by Deligiannis *et al.* (2011) and Sanford *et al.* (2009) respectively. Higher value (118.4 mgKOH/g) was reported by Caetano *et al.* (2012) on WCR oil which is much greater than the acid value of this study.

The density (mass per unit volume) of the waste coffee residue oil was found to be 0.9228 g/cm³. This value is lower than the value (0.9338 g/cm³) reported by Deligiannis *et al.* (2011) and higher than the value (0.917 g/cm³) reported by Caetano *et al.* (2012). The density of WCR oil exceeds the EN standard specification of 0.86-0.9 g/cm³. This high density could be due to the unsaturation level in WCR oil. Thus, the oil cannot be used directly as fuel; since this high density would polymerise and leads to formation of deposits in car engine.

Table 4.2: Characterization of the oil extracted from WCR

Property	Units	Test Methods	Limits	Results
Density (15 °C)	g/cm³	D1298	0.86-0.9 g/cm³	0.9228
Kinematic Viscosity(40 °C)	mm²/s	D 445	1.9 – 6.0	35.50
Gross calorific value	MJ/kg	D 240	Report	37.88
Cloud point	°C	D 97	Report	10
Iodine Value	gI2/100g	EN14111	120 max	75.94
Water and sediment	%volume	D2709	0.050 max	0.025
Acid value	Mg	D974	0.8 max	14.65
Flash point	°C	D93	130 min	>300
Saponification value	mgKOH/g	AOCS Cd 3-25	--------	175.75
Peroxide value	Meq/kg	D3703	---------	2.99

Viscosity is a key fuel property because it influences the atomization of a fuel upon injection into the diesel engine ignition chamber and ultimately, the formation of engine deposits (Knothe and Steidley, 2005). The viscosity of the WCR oil was found to be 35.5 mm²/s (Table 4.2). This value was higher than 22.23 mm²/s reported by Caetano et al. (2012). But lower than the values 46.965 mm²/s and 40.97 mm²/s reported by Deligiannis *et al.* (2011) and Sanford et al. (2009) respectively. The result of this study is much higher than the ASTM and EN standard specifications of 1.9-6.0 and 3.5-5.0 mm²/s respectively. The more viscous oil is, the better its use as a lubricant; hence WCR oil will have high lubricating properties (Olaniyan and Oje,

2007). The viscosity of the WCR oil must be reduced for biodiesel application because high viscosity of vegetable oil is not suitable. If it is used directly as engine fuel, often it results in operational problem such as carbon deposit, oil ring sticking, thickening and gelling of lubricating oil. Different methods such as pre heating, blending, ultrasonically assisted methanol transesterification, supercritical methanol transesterification and acid or base catalyzed transesterification are being used to reduce the viscosity and make them suitable for engine application (Adeyinka *et al.*, 2011; Banapurmath, 2008; Pramanik, 2003).

Moisture in vegetable oils is a great impediment to the formation of esters due to increase in tendency of soap formation and thus will have to be minimal for transesterification to occur (Abdulkareem *et al.*, 2011). In this study, the moisture content of the oil was found to be 0.025% which is lower than the 0.05% maximum limit of the ASTM standard. This value shows appreciable consistency with the literature (Sanford *et al.*, 2009). The low water content confirms that the oil is of good quality and could not be easily subjected to contamination/rancidity (Fellows; 1997). The presence of much water in biodiesel feedstock has negative impacts on transesterification process and the effective cost of the feedstock (Ma and Hanna; 1999). This is because it supports microbial growth.

Gross heat content (Hg) of the WCR oil was determined to be 37.88 MJ/kg. The relatively high Gross calorific value suggests that this oil can be used for direct combustion. The value of this work is lower than the value (39.49 MJ/kg) reported by Deligiannis *et al.* (2011) and higher than 36.4 MJ/kg reported by Caetano *et al.* (2012) on calorific value of WCR oil.

The iodine value shows the level of unsaturation of the oil and also influences the oxidation and deposition formed in diesel engines. Iodine value obtained for WCR oil was seen to be 75.94 gI2/100g and the value fall within the acceptable ASTM limit as shown in Table 4.3. Oils with iodine value above 125 are classified as drying oils; those with iodine value 110–140 are classified as semidrying oils. Those with iodine value less than 110 are considered as nondrying oil (Nkolika, 2012). Thus, the WCR oil is classified as non-drying oil since its iodine value is below 110 gI2/100g. According to Kochhar (1998) oils that are not susceptible to drying are also not susceptible to becoming rancid. This implies that WCR oil can be preserved for a long period of time. If used as straight vegetable oil, it is not more likely to polymerize in the heat of the engine. As a result, WCR oil is suitable for biodiesel production.

Cloud point of WCR oil was found to be high (10 °C) making the oil unsuited for production of biodiesel. Thus, biodiesel from this oil would not be usable in cold countries without the use of additives to improve their cold filter plugging points (CFPP). The result found in this study was lower than the value (13 °C) reported by Deligiannis *et al.* (2011).

Flash point of crude oil of WCR was >300 °C which is very high making it better suited for biodiesel production with respect to safety during storage and transportation (Table 4.2). It was far above the 93 °C minimum ASTM standard recommended range and therefore poses no risk of fire outbreaks in case of accidents.

The saponification value (SV) indicates the ability of the oil to make soap during the transesterification reaction. The saponification value of the WCR oil analyzed was found to be 175.75 mg/KOH/g (Table 4.2). This high saponification value indicates an index of high average molecular weight of triacylglycerols in the oil. It also indicates the presence of high percentage of free fatty acids in the oil (Omolara and Dosumu, 2009) and thus implies the possible tendency to soap formation and difficulties in separation of products if utilized for biodiesel production. This would also suggest that using the oil for biodiesel production would lead to very low yield in the methyl ester. Therefore, this oil may be a good raw material for soap making, but it needs pretreatment prior to using it for biodiesel production.

The peroxide concentration, usually expressed as peroxide value (PV), is a measure of oxidation or rancidity in its early stages (Gunstone, 2004). Crude WCR oil was found to have a peroxide value of 2.99 Meq/kg. This low value attests to the oxidative stability of the oil.

4.3. Biodiesel yield of WCR oil

The WCR oil was first subjected to acid catalyzed esterification step at 60 °C using H_2SO_4 (10 wt% of total fatty acids content) as a catalyst for 2 h reaction time to reduce the high free fatty acid content. The %FFA of the WCR oil was decreased from 7.33% to 0.9% after three consecutive esterification steps. After esterification, the oil pretreatment loss was found to be 8.67% based on the initial sample of WCR oil. Transesterification of the pretreated WCR oil was then carried out using NaOH as a catalyst and methanol to oil 6:1 molar ratio at 60 °C for 2 h. The biodiesel yield relative to WCR oil and acid pretreated WCR oil was found to be 73.4 and 80.4 wt%, respectively. The yield is lower than those reported for WCR biodiesel (100% and 92%) by Kondamudi *et al.* (2008) and Deligiannis *et al.* (2011), respectively. However, the yield obtained is higher than the reported values (58.8-62.2 wt%) by Caetano *et al.* (2012). The difference in the biodiesel yield could possibly be due to the variation in the free fatty acid content of the feedstock.

4.4. Waste Coffee Residue methyl ester Fuel properties

The biodiesel produced from WCR oil was analyzed for its fuel properties and compared with ASTM D 6751 and EN 14214 standards (Table 4.3). The fuel properties of WCR biodiesel meet all the quality standards set (ASTM D6751) except for the acid value (Table 4.3).

4.4.1. Density (15 °C)

The standard for biodiesel states that the fuel should have a density between 0.86 and 0.90 g/cm^3 (EN 14214). This property is important mainly in airless combustion systems because it influences the efficiency of atomization of the fuel (Ryan et al., 1984). The result obtained for WCR biodiesel was 0.8915 g/cm^3 which is in agreement with the EN standard specification for density. The value obtained is also comparable with those reported for density of WCR biodiesel by Caetano *et al.* (2012), 0.911 g/cm^3, Deligiannis *et al.* (2011), 0.8943g/cm^3 and Sanford *et al.* (2009), 0.8815 g/cm^3.

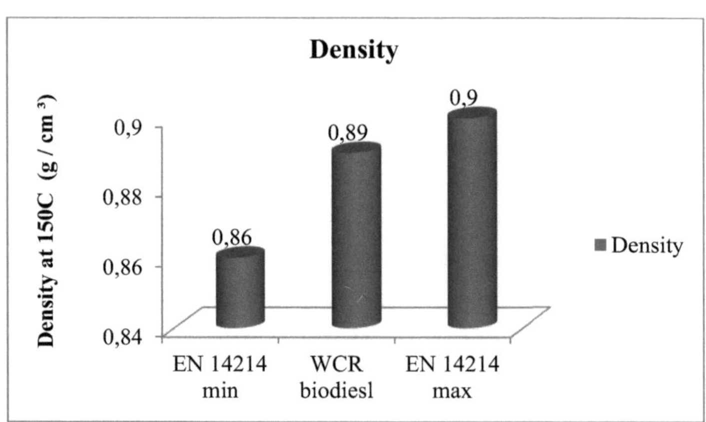

Fig.20: Density of WCR biodiesel and EN 14214 standard specification

4.4.2. Kinematic viscosity (40 °C)

The kinematic viscosity (measurement of internal resistance of liquid flow) is an important parameter for the vehicle injection system and fuel pumping system. According to ASTM D 6751 and EN 14214 for biodiesel to be used in diesel engines, the kinematic viscosity must be between 1.9 and 6.0 mm^2/s and 3.5 and 5.0 mm^2/s respectively. The kinematic viscosity of the biodiesel produced from WCR oil was 5.3 mm^2/s and met the ASTM standard, but it exceeds to the EN 14214 standard specification of 3.5 -5.0 mm^2/s (fig.21). They are different values for kinematic viscosity of WCR biodiesel reported in the literature, 4.852 mm^2/s, 5.84 mm^2/s, 5.612 mm^2/s, and 12.88 mm^2/s reported by Sanford *et al.* (2009), Kondamudi *et al.* (2008), Deligiannis *et al.* (2011) and Caetano *et al.* (2012) respectively. Variation of values could be due to the different fatty acids composition of a given biodiesel sample. It increases with increasing length of fatty acid chain (Knothe, 2002).

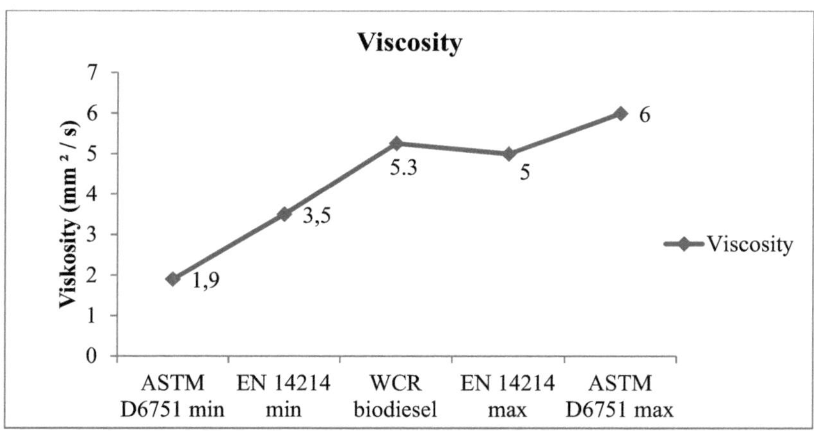

Fig. 21: Viscosity of WCR biodiesel in comparison with ASTM and EN standards

4.4.3. Gross calorific value

Calorific value (CV) is another important property of biodiesel as it is aimed for use as diesel fuel substitute. Calorific value of WCR biodiesel was found to be 38.4 MJ/kg. No significant change in heating value was observed after transesterification of the WCR oil (37.88 MJ/kg) and confirms the report of Lague *et al.*, (1987) that CV is not affected by transesterification. The CV of WCR biodiesel (38.4 MJ/kg) was found to be lower than that of diesel. This could be due to the oxygen content of biodiesel which improves combustion process and decreases its oxidation potential. The obtained result was higher than the value of 36.4 MJ/kg reported by Caetano *et al.* (2012) and lower than 39.49 MJ/kg reported by Deligiannis *et al.* (2011). The CV difference could be due to the composition of the fatty acids, where grown and its vintage among other factors.

4.4.4. Cloud point

Cloud point (CP) is used to measure the cold temperature usability of ester as fuel. Cold temperature behavior of biodiesel is an important quality criterion, as frozen fuel may cause blockage of the fuel lines and filters and starve the engine of fuel. The cold temperature usability of WCR biodiesel was found to be 14 °C, which was higher than corresponding WCR oil. The obtained result was higher than those reported by Kondamudi *et al.* (2008) and Deligiannis *et al.* (2011), 11°C and 13 °C respectively. The cloud point of 0.2 °C reported by Sanford *et al.* (2009) was much lower than this study and other works on WCR biodiesel. Higher cloud point of the WCR biodiesel in this study may be due to the presence of some unknown compounds in WCR and formation of polymerized esters during transesterification. From this result it is clear that WCR biodiesel would be a suitable candidate as a diesel fuel substitute in tropical countries and not in colder climate conditions. However, blending and winterization of WCR ester or some additives may be necessary for the application of the ester in cold temperature.

4.4.5. Acid value (AV)

The acid value of WCR biodiesel was determined according to ASTM D974 and was found to be 0.78 mmgKOH/g (Table 4.3). Although the acid value of WCR biodiesel was decreased from 14.65 to 0.78 mgKOH/g after esterification and transesterification, still exceeds the maximum recommended standard value of 0.5 mgKOH/g for latest ASTM D 6751 and EN 14214 standard specifications. This was may be due to the tendency of the methyl ester of the oil to hydrolyze rendering more free fatty acids. The value of this work is better than 2.14 mgKOH/g reported by Caetano *et al.* (2012). Lower values of 0.36 mgKOH/g and 0.35 mgKOH/g were reported by Deligiannis *et al.* (2011) and Kondamudi *et al.* (2008) respectively. High acid values on the feedstock indicates unrefined or poorly refined product oil source due to poor process control.

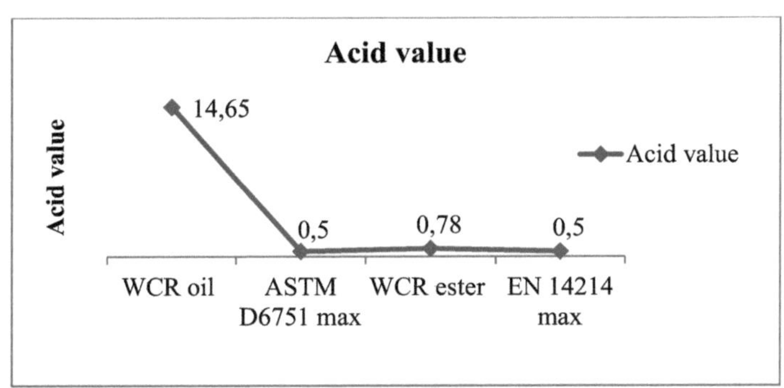

Fig. 22: Acid value of WCR oil, biodiesel, ASTM and EN standards

Table 4.3: Quality Characterization of the WCR Biodiesel

Property	Units	Test Methods	Limits ASTM6751	Limits EN 14214	WCR oil methyl ester
Density (15 °C)	g/cm³	D1298	--------	0.86 – 0.9	0.8915
Kinematic viscosity(40 °C)	mm²/s	D445	1.9 -6.0	3.5-5	5.26
Gross calorific value	MJ/kg	D240	--------	-------	38.4
Cloud point	°C	D97	Report	-------	14
iodine value	gI2/100g	EN14111	--------	120 max	73.41
water and sediment	%volume	D2709	0.05 max	500 mg/kg	<0.01
Ash content	W%	D874	0.02 max	0.02 max	0.0123
acid value	mgKOH/g	D664	0.5 max	0.50 max	0.78
Carbon residue	% mass	D189	0.05 max	0.3max	0.033
flash point	°C	D93	93.0 min	120 min	222
Copper corrosion	Max.	D130	No. 3 max	No. 1 max	1a
Cetane number	Min.	D613	47	51	59
Distillation, 90% recovery	°C	D86	360°Cmax	-----	349

4.4.6. Cetane index/ number

Cetane index is used to evaluate combustion quality of esters. The higher the cetane index, the better the quality of esters as diesel fuel. Cetane index of the biodiesel prepared from WCR was found to be 60.84. According to Patel (1999) correlation, the cetane number of WCR methyl ester was estimated to be 59.34, which well exceeds the minimum cetane number of 47 and 51 prescribed in the ASTM D6751-09 and EN 14214 standard specifications for the cetane number respectively (fig.23). A higher cetane number of WCR methyl ester shows that the fuel is more ignitable. References (Giwa *et al.*, 2010, Rashid *et al.*, 2009 and Usta *et al.*, 2005) reported that the cetane numbers of biodiesel produced from oils with analogous FA profile such as soybean, egusi, and sunflower were 53.66, 54 and 55, respectively.

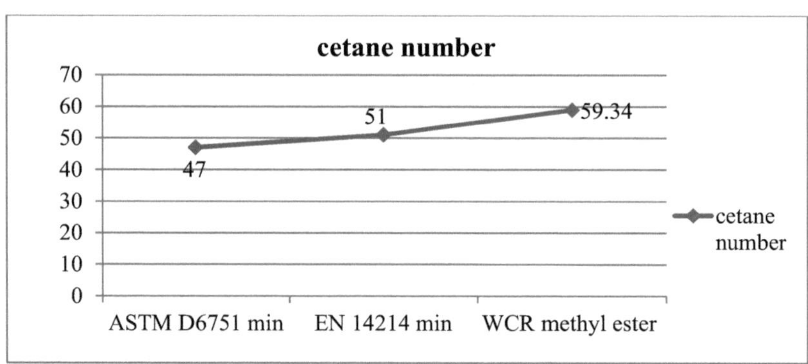

Fig. 23: Cetane number of WCR biodiesel in comparison with ASTM and EN standards

4.4.7. Iodine Value (IV)

The Iodine Value, (a measure of total unsaturation) of the WCR biodiesel was found to be 73.41 gI2/100g. This result was slightly lower than the value of WCR oil (75.94 gI2/100g) before transesterification (fig.24).The limitation of unsaturation of fatty acid is vital due to the fact that heating highly unsaturated fatty acids results in polymerization of glycerides which could lead to the formation of deposits (Mittlebach, 1996). According to the EN 14111 this value should be lower than 120 gI2/100g. The obtained value implies biodiesel produced from WCR oil has iodine value within the EN 14214 specification, thus has a moderate degree of unsaturation. As a result, the biodiesel from WCR oil is suitable to be used as an alternative fuel.

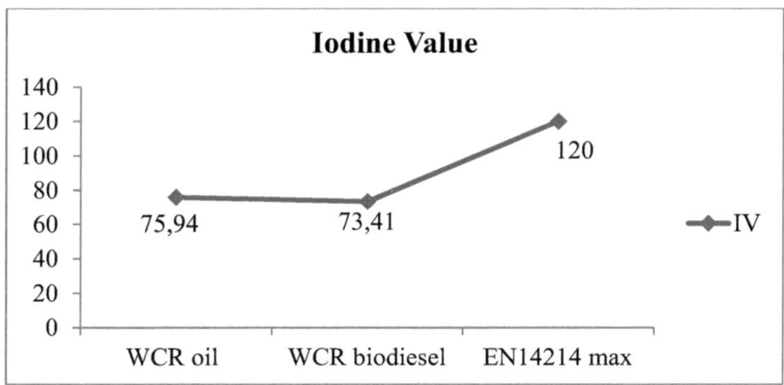

Fig. 24: Iodine value of WCR oil, biodiesel, and EN standard specification

4.4.8. Water and sediment

Biodiesel contaminated with water can cause an unwanted reaction, producing free fatty acids, growth of microorganisms, corrosion and malfunctions of the engine (Benjumea *et al.*, 2008). Due to those negative effects, ASTM D 6751and EN 14214 set the maximum allowable content of 0.05% for water in biodiesel. Water and sediment test resulted in value of <0.01% for the biodiesel produced from WCR oil. This value was well below the maximum limitation of both ASTM D 6751 and EN 14214 biodiesel standards. This can be attributed to the dry wash and drying process. The obtained value was lower than the value of <0.05% reported by Sanford *et al.* (2009) for WCR biodiesel.

4.4.9. Distillation Temperature

Distillation temperature was determined as per the ASTM D86. The data tested for the methyl ester of WCR includes the initial boiling point (IBP), boiling temperature corresponding to increments of the volume of fuel distilled, final boiling point (FBP) and % total recovery (see appendix 5). The initial and final boiling points of WCR biodiesel were found to be 314.5 and 362.5 °C respectively.

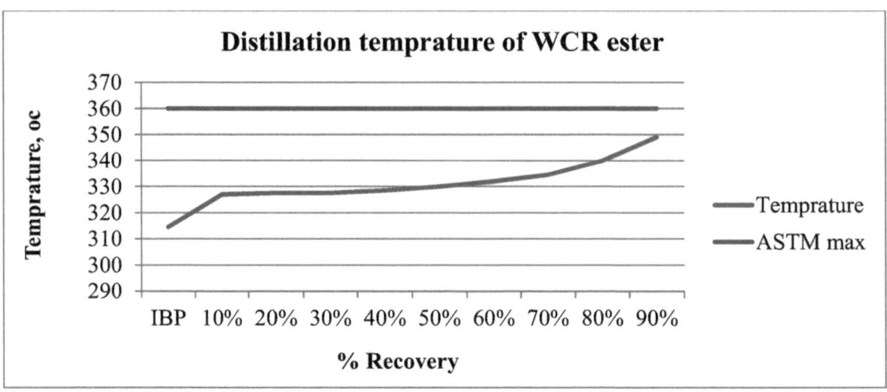

Fig.25: Distillation temperature of WCR biodiesel in relative to ASTM specification

WCR biodiesel at 90 v% recovered boiling temperature was determined 349 °C. This value was in agreement with ASTM specification of 360 °C max (fig.25). WCR Biodiesel exhibits a narrow boiling range compared to the distillation range of diesel. This is because fatty acid chains in the WCR biodiesel produced was mainly comprised of straight chain hydrocarbons with 16 to 18 carbons that have similar boiling temperatures (Chevron, 2006).This signifies the biodiesel produced is likely to be less volatile than that of the conventional No.2 diesel that has the initial boiling point of 186 °C. Cracking with smoke formation was observed at >95 v% temperatures for WCR biodiesel.

4.4.10. Ash content

The ash content determination of biodiesel sample in this study was done following ASTM D874. Ash forming materials especially the unremoved catalyst used in biodiesel production can cause engine deposits and contribute to wear and tear at the fuel injector, fuel pump, piston and ring. Thus, there is a need to limit the ash content in the biodiesel sample. Currently, a maximum of 0.02 wt% is stipulated in the biodiesel specification of ASTM 6751 and EN 14214. The ash content obtained for WCR biodiesel was 0.0123 wt%. The value obtained was well below the ASTM and EN standards of 0.02 wt% max. (fig.26). This indicates that it may likely have lower mineral contents which otherwise would contribute to some level of air pollutants like SOx and

NOx. Therefore, WCR emissions from exhaust of vehicles will help to reduce the pollution introduced to the atmosphere compared to that of fossil diesel.

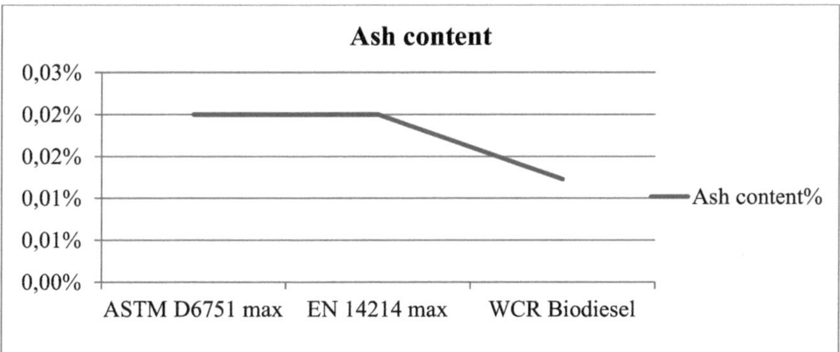

Fig.26: WCR biodiesel ash content in comparative to ASTM and EN specifications

4.4.11. The Conradson Carbon residue

The conradson carbon residue was performed according to the standard procedure of ASTM D189. It is an indication of the amount of carbon residue that will be deposited in the engine as the result of fuel combustion. The Carbon residue of WCR biodiesel was found to be 0.033 wt%. This value was well below the ASTM D6751 specification of 0.05% mass max. This value indicates a low level of glycerine. Thus WCR biodiesel carbon has low scuffing on metal surfaces and a buildup of carbon around the injector tips could slightly interfere with the spray pattern and amount of fuel delivered. Sanford *et al.* (2009) reported high value (0.4%) on WCR biodiesel.

4.4.12. Flash point

The flash point was measured with a Pensky-Martens closed cup tester using ASTM D93. The flash point temperature of a fuel is the minimum temperature at which the fuel will ignite or flash when an ignition source is applied. The flash point of pure biodiesels is usually higher than the ASTM limits, but fall rapidly with increasing amount of methanol. The flash point of the WCR biodiesel was found to be 222 °C, which is higher than the ASTM limits for biodiesel (Fig.27). This high flash point of WCR biodiesel leads to its safer handling and storage. Deligiannis *et al.*

(2011) and Sanford *et al.* (2009) reported >120 and >160 °C flash points of WCR ester, respectively.

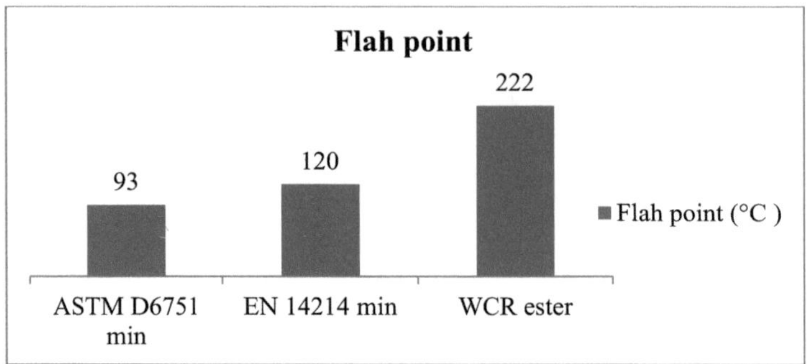

Fig.27: WCR biodiesel flash point in relation to ASTM and EN standard specifications

4.4.13. Copper corrosion

The copper strip corrosion is used for the detection of the corrosiveness to copper of biodiesel fuels, which can lead to problems of corrosion in storage tanks and some engine parts. The copper strip corrosion test result for WCR methyl biodiesel was No.1a, which is the lowest level of corrosiveness. This detected result met the acceptable specific standards stipulated by both EN 14214 (No. 1a max) and ASTM D6751 (No. 3 max). Thus, corrosion would not be a problem for WCR biodiesel.

4.5. Fatty acid composition of WCR biodiesel

The fatty acid composition of WCR oil methyl ester was determined by gas chromatography (GC). GC analysis showed the presence of C16-C18 and two unidentified methyl ester of fatty acids (see appendix 3). WCR biodiesel consists of both saturated and unsaturated methyl esters (fig. 28), where more than 97.7% of the total composition was methyl esters of linoleic acid (39.8%), palmitic acid (37.6%), oleic (12.7%) and stearic acid (7.6%). Linoleic acid showed the highest percentage of composition followed by palmitic acid. Apart from the biodiesel, this linoleic– palmitic fatty acid dominance of WCR implies, it can be used in the beauty products industry. Kondamudi *et al.* (2008) also reported a similar composition on WCR biodiesel.

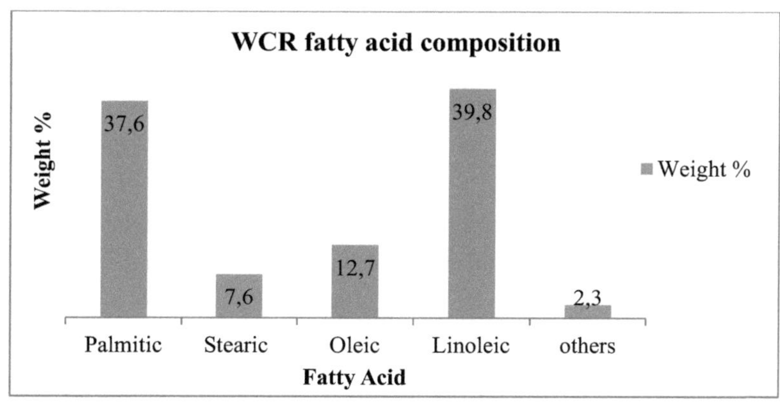

Fig.28: WCR methyl ester fatty acid composition

4.6. Bioethanol from solid waste remaining after oil extraction of WCR (Spent of WCR)

The spent of WCR after oil extraction was analyzed for its ethanol content. Hydrolysis of the spent with dilute H_2SO_4 and subsequent fermentation with yeast yielded ethanol.

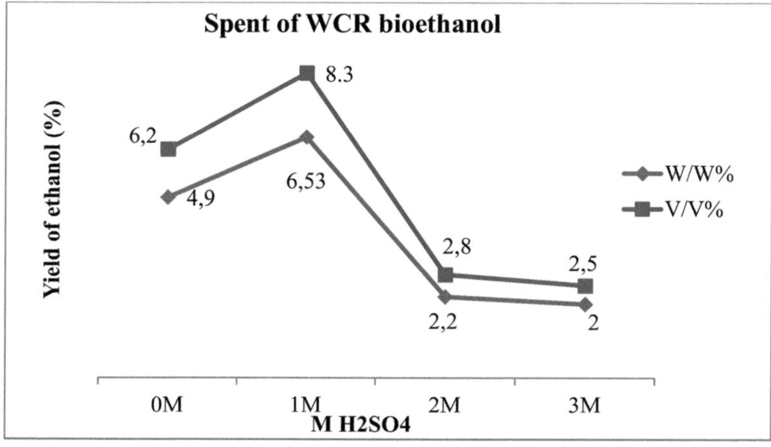

Fig. 29: Bioethanol yield (%) of the solid waste after oil extraction (spent of WCR)

Dilute sulfuric acid (each of 1, 2 and 3 M H_2SO_4) were used for hydrolysis, which resulted in bioethanol yield (% *v/v*) of 8.27, 2.8 and 2.5 respectively (see appendix 6). The solid to liquid ratio of 1:10 hydrolyzed by distilled water was resulted 6.2 % *v/v* of bioethanol yield. The maximum bioethanol yield (8.3% *v/v*) was achieved at 1M of sulfuric acid (Fig. 29).

Although, the bioethanol yield is increasing from Hydrolysis with distilled water to 1M of sulfuric acid, the obtained result is more comparable. The reason that lower bioethanol yield obtained with distilled water hydrolysis in comparison to acid hydrolysis could be due to the lignocellulose content of the spent of WCR that couldn't depolymerised in to sugars with water. The carbohydrate fraction (hemicellulose and cellulose) of spent of WCR can be depolymerised into sugars which act as a primary carbon source for the microbial biocatalysts for the production of ethanol by acid hydrolysis (Carvalheiro *et al.* 2008; Mussatto and Teixeira 2010). Further increase in molarity of sulfuric acid beyond 1M of sulfuric acid resulted low bioethanol yield. This decrease in bioethanol yield may account for the further sugar degradation that occurred under the severe acidity. In general, this result indicates that extreme acidity had an adverse effect on sugar conversion of spent of WCR (Nutawan *et al.*, 2010). The obtained result of this study (65.3 g/kg) was lower than the result (110g/kg) reported by Caetano *et al.* (2011) on the original WCR (before oil extraction) without optimization. He also reported that the result could be 200 g/kg if fermentation optimization is applied on the original WCR. Spent of WCR has high ethanol yield compared to the waste materials reported by Yamada *et al.* (2011) on rice straw (7.5 g/l), Dhabekar and Chandak (2010) on beet waste (2.15%) and banana peels (1.90%), and Quevedo- Hidalgo *et al.* (2013) on Chrysanthemum waste degradation (0.45 wt%). Therefore, this result shows that the spent of WCR has still the potential as source of bioethanol.

4.7. Solid fuel and compost from the WCR after Bioethanol production

Analysis of the solid waste remaining after Bioethanol production of spent of WCR for solid fuel and compost was carried out. Table 4.4 below showed the result of the quality parameters analyzed.

Table 4.4: characteristics of WCR after bioethanol production for solid fuel and compost

property		Units	Solid waste after Bioethanol
Gross calorific value		MJkg-1	20.8
Proximate Analysis	Fixed Carbon	w %	17.59
	Moisture	w %	4.5
	Ash	w %	2.04
	Volatile matter	w %	75.87
Total nitrogen		w %	2.30
Total carbon		w %	50.43

All the % values are on a dry weight basis, except the moisture content.

4.7.1. Solid waste remaining after Bioethanol production for solid fuel

Measurement of the calorific value of WCR after Bioethanol production provided the result of 20.8 MJ/kg (expressed per dry weight of solid) (table 4.4.). According to the obtained result of this study, the calorific value of WCR is only marginally affected by bioethanol production as with reference to the result of Deligiannis *et al.* (2011) which was 21.16 MJ/kg from solid waste remaining after oil extraction. This signifies the energy released by its combustion is very large. In particular, it appears to be in the highest range of those of sawdust and most biomass wastes and residues, which are roughly in the range of 15-21 MJ/kg (Demirbas, 1997; Poskart and Szecówka, 2008). It has also high calorific value compared to the conventional biomass reported by Hoi (2003) like bagasse (7.7-8 MJ/Kg), rice husks (14 MJ/Kg), coffee husk (16 MJ/Kg) and wood (8.4- 17).

Determination of the fuel combustion properties was also conducted based on the proximate analysis of WCR after bioethanol production. The ash and moisture content of the solid waste was low, whereas its volatile mater was high (Table 4.4). According to Quaak *et al.*, (1999) fuels should have ash content less than 5%.

The presence of moisture in fuel lowers its effective heating value since a portion of the heat of combustion is utilized in evaporating the contained moisture (Quaak *et al.*, 1999). Solid waste in this study has been found to have low moisture (4.5%).The result obtained is comparable with those reported in the literature and thus, the spent of WCR after bioethanol production can be used as direct combustible biofuel.

Volatile matter burns as visible flame on supply of sufficient air, time, temperature and turbulence (Lamare and Wing, 2001). The volatile matter of the spent in this study was found to be high (75.87%). Pate and Gami, (2012) reported that if volatile matter in the fuel is higher, then large amount of secondary air with high pressure need to be supplied at strategic location for effective combustion.

The fixed carbon also affects the heating value of the fuel in such a way that the higher the fixed carbon, the higher the thermal value of fuel. In general, an increase in the ash content corresponds to a decrease in the fixed carbon content and hence a decrease in heating value (Unesco, 1988). As discussed above this study has low ash content which contributes to high fixed carbon. As a result, the thermal value of WCR after bioethanol production was high.

4.7.2. Solid waste remaining after Bioethanol production for compost

Due to the diverse nature of feed stock and composting processes, the quality of available compost materials can vary widely. Carbon to nitrogen ratio is a parameter used to determine if compost is nitrogen stable. According to EPA (2008), Ideal waste coffee residues for the soil need a carbon to nitrogen ratio (wt) of 20:1. The carbon to nitrogen ratio after the bioethanol production   process was found to be 21.92:1 (table 4.4). This C/N ratio of solid waste after bioethanol production of spent of WCR met the ideal carbon to nitrogen ratio reported by EPA (2008). B.C. Agriculture and Food, (1996) also confirms that a C: N ratio of between 20 and 40:1 is often suitable for composting depending on the make-up of the feedstock. Caetano *et al.* (2012) reported that the carbon to nitrogen ratio of the original waste coffee residue (24.9:1). Standing from this literature (Caetano *et al.* (2012) the value determined in this study has no significant change in the C/N ratio of the same material passing oil extraction and bioethanol production processes. This indicated that the processed coffee residues could still be used as compost for the garden.

5. CONCLUSION AND RECOMMENDATIONS

5.1. Conclusion

The aim of this study was to evaluate waste coffee residue remaining after brewing coffee as a potential alternative feedstock for biodiesel production. Around 19.73 %*w/w* of the Waste coffee residue oil was extracted via hexane and physicochemical characterization was carried out. The obtained WCR oil was chemically converted to fatty acid methyl ester via two-step reaction process, acid catalyzed esterification process and followed by base catalyzed transesterification process due to its high FFA content (7.33 %). The first acid-catalyzed esterification step, carried out at 60 ºC for 2 h at 20:1 molar ratio of methanol to oil and 10 % sulfuric acid concentration (*w/w*, based on FFA), reduced the FFA level from 7.33 % to 0.9 % . The second alkali-catalyzed transesterification step, conducted at 60 ºC for 2 h at 6:1 molar ratio of methanol to oil and 1 %*w/w* sodium hydroxide (based on pretreated oil), was converted the esterified waste coffee residue oil into biodiesel with 80.4 % yield. The waste coffee residue oil biodiesel produced was characterized for its viscosity, density, flash point, acid value, cloud point, distillation temperature, ash content, carbon residue, calorific value, water and sediment, Copper corrosion and iodine value, and has fuel properties that met the latest ASTM D 6751 standard specification except for acid value. GC analyses indicated that the waste coffee residue biodiesel comprised of saturated (45.2 %), unsaturated (52.5 %) and two unidentified (2.3 %) esters.

Furthermore, the potential of solid waste remaining after oil extraction (spent of WCR) for bioethanol production was carried out by hydrolysis of a solid to liquid ratio of 1:10, using dilute sulfuric acid and distilled water followed by fermentation, distillation and finally the ethanol concentration analyzed by FTIR. The bioethanol production from spent of WCR test have shown that dilute acid is preferable than distilled water hydrolysis. The result of the study showed that the highest bioethanol yield of 8.3 %*v/v* was obtained with 1 M H_2SO_4, which is significant.

Finally, the quality property of solid waste after bioethanol production, including ash, moisture, volatile matter, fixed carbon, calorific value, total nitrogen and total carbon, were characterized and found to be it still has the potential to use for solid fuel and compost. It has high calorific value compared to the conventional biomass like bagasse, rice husks, sawdust, coffee husk and wood. The carbon to nitrogen ratio after the bioethanol production   process (21.9:1) of the solid waste makes it suitable as compost raw feedstock.

Therefore, as a cheap feedstock, waste coffee residue can be potentially used as a raw material for biodiesel and bioethanol production on a commercial scale. This can supplement the Ethiopian energy sector approximately by 13.3 and 0.41 million gallons of biodiesel and bioethanol respectively by sequential utilization of the same WCR per year. This work may present new coming to sequential production of biofuels from waste materials.

5.2. Recommendations

- Further investigation of WCR biodiesel in Clarifying the uncertainties regarding oil extraction from WCR, such as solvent selection *vs* extraction outcome, influence of solvent on oil composition, optimal extraction time, etc.; should be conducted.
- Further investigation should be done to analyze the oil content of the WCR from different coffee Arabica species of Ethiopia.
- An economic feasibility analysis of the overall conversion process of biodiesel and bioethanol of WCR is necessary for the purpose of commercialization.
- The biodiesel quality parameters including acid value, carbon residue, distillation temperature at 90 v% recovered and sulfated ash were tested based on ASTM D975 standard. These and others uncovered because of the absence of the testing technology (oxidation stability) in the study should have to be tested based on ASTM and EN standards for biodiesel.
- Detail investigation should be carried out on the sequential utilization of WCR for biodiesel, bioethanol, solid fuel and compost
- Optimization of the bioethanol production (hydrolysis and fermentation) of the waste solid after oil extraction (spent of WCR) should be conducted
- Comparison study on the original WCR and spent of WCR on bioethanol production should be further studied to realize its feasibility

REFERENCES

Abdulkareem, A.S., Uthman, H., Afolabi, A.S., Awonebe, O.L. (2011).Extraction and Optimization of Oil from Moringa Oleifera Seed as an Alternative Feedstock for the production of Biodeisel. Majid, N., Mostafa, K., editors: *Sustainable Growth and Application in Renewable Energy Sources:* InTech; Pp243-268.

Abreu, F.R., Lima, D.G., Hamu, C.W., Suarz, P.A.Z. (2004). Utilization of metal complexes as catalysts in the transesterification of Brazilian vegetable oils with different alcohols: *J. Mol. Catal. A: Chem.*, 209: 29-33.

Abu Tefera (2012). Ethiopia Coffee Annual Report: global agricultural information network.

Alamu, O. J., Waheed, M. A., Jekayinfa, S. O. (2008). Effect of ethanol–palm kernel oil ratio on alkali-catalyzed biodiesel yield: *Fuel* 87: 1529–1533.

Alamu, O.J, Waheed, M.A., Jekayinfa, S.O., Akintola, T.A. (2007). Optimal transesterification duration for biodiesel production from nigerian palm kernel oil. *Agric Eng Int: CIGR Ejournal*; IX.

Alberta Agriculture, Food and Rural Development (2006). Vegetable Oil Fuels. http://www1.agric.gov.ab.ca/$department/deptdocs.nsf/all/eng4435

Alemayhu Tegegn (2007). High level African Biofuel Seminar AU. Ministry of Mines and Energy, Addis Ababa, Ethiopia.

Al-Hamamre, Z., Foerster, S., Hartmann, F., Kroger, M., and Kaltschmitt, M. (2012). Oil extracted from spent coffee grounds as a renewable source for fatty acid methyl ester manufacturing, *Fuel,* vol. 96, pp. 70-76.

Allen, C.A.W., K.C. Watts, R.G. Ackman, and M.J. Pegg, (1999). Predicting the Viscosity of Biodiesel Fuels from Their Fatty Acid Ester Composition; *J. Ener. Fuel* 78; 1319–1326

ASTM (2009). Standard Specification for Biodiesel Fuel (B100) Blend Stock for Distillate Fuels: 1131-1136.

ASTM D 1250-80. (1980). Petroleum Measurement Tables. Volume Correction Factors; *American Petroleum Institute*: Vol. 5, Tables 23B, 24B. USA.

ASTM Standard D1298, 1999 (2005), "Standard Test Method for Density, Relative Density (Specific Gravity), or API Gravity of Crude Petroleum and Liquid Petroleum Products by Hydrometer Method," ASTM International, West Conshohocken, PA.

ASTM Standard D2500, (2005). "Standard Test Method for Cloud Point of Petroleum Products:" ASTM International, West Conshohocken, PA.

ASTM Standard D6751, (2009). "Standard Specification for Biodiesel Fuel Blend Stock (B100) for Middle Distillate Fuels": ASTM International, West Conshohocken, PA.

ASTM Standard D93, (2008). Standard Test Methods for Flash Point by Pensky- Martens Closed Cup Tester, ASTM International, West Conshohocken, PA.

ASTM Standard specification for biodiesel fuel (B100) blend stock for distillate fuels (2008). In: Annual Book of ASTM Standards, ASTM International, West Conshohocken, Method D6751 08.

Atadashi, I.M.; Aroua, M.K. and Abdul Aziz, A. (2011). Biodiesel separation and purification: A review. *Renewable Energy* Volume 36, Issue 2, Pages 437–443.

Ayele kefale, Mesfin Redi and Araya asfaw (2012). Bioethanol production and optimization test from agricultural waste: the case of wet coffee processing waste (pulp): *international journal of renewable energy research*, vol.2, No.3.

Azam, M. M., Waris, A., Nahar, N. M. (2005). *Biomass Bioenergy 29*, 293–302.

B.C. Agriculture and Food (1996). Composting fact sheet. British Columbia Ministry of Agriculture and Food, Abbotsford, BC, Canada.

Banapurmath, N.R., Tewari P.G., Hosmath, R.S. (2008). Performance and emission characteristics of a DI compression ignition engine operated on Honge, Jatropha and sesame oil methyl esters. *Renewable Energy* 33: 1982–1988

Benjumea, P., Agudelo, J., Agudelo, A. (2008). Basic properties of palm oil biodiesel-diesel blends. *Fuel;* 87 2069-2075.

Bernardes, O.L., Bevilaqua, J.V., Leal, M.C.M.R., Freire, D.M.G., Langone, M.A.P. (2007). Biodiesel fuel production by the transesterification reaction of soybean oil using immobilized lipase. *Appl Biochem Biotechnol;* 137–140:105–14.

Bkansedo, J. (2009). Synthesis of biodiesel from palm oil and sea mango oil using sulfated zirconia catalyst. Universiti sains Malaysia.

Bondioli, P., Gasparoli, A., Bella, L. D., Tagliabue, S., Toso, G. (2003). Biodiesel stability under commercial storage conditions over one year. *Eur. J. Lipid Sci. Technol.* 105: 35–741.

Bureau of African Affairs (2012). Background Note: Ethiopia. U.S. Department of state.

Burtis, B. (2006). Biofuels and Energy Production Options for the Farm. Climate Change and Northeast Agriculture: http://www.agric.wa.gov.au/pls/portal30/docs/FOLDER/IKMP/SUST/BIOFUEL/Econo m csBiofulProdn.pdf .

Caetano, N. S., Silva, V.F.M. Mata,T. M. (2012). Valorization of Coffee Grounds for Biodiesel Production: *the Italian Association of Chemical Engineering*, VOL. *26*.

Caetano, N.S. and Mata, T.M. (2011). Residual Biomass utilization for biofuel production. *Jornadas Tecnicas Internacionais De Resdues.*

Calixto, F. , Fernandes, J. , Couto, R. , Hernandez, E. J. , Najdanovic-Visaka, V., and Simoes, P. C.(2011).Synthesis of fatty acid methyl esters via direct transesterification with methanol/carbon dioxide mixtures from spent coffee grounds feedstock," *Green Chemistry,* vol. 13, pp. 1196-1202.

Campo, P., Zhao, Y., Suidan, M. T., Venosa, A. D., Sorial, G. A. (2007). *Chemosphere* 68, 2054–2062.15

Canakci, M. and Gerpen, J.V. (2001). "Biodiesel Production from Oils and Fats with High Free Fatty Acids: *American Society of Agricultural Engineers*, vol. 44, No. 6, pp.1429-1436.

Canakci, M. and Gerpen, J.V. (2003). A pilot plant to produce biodiesel from high free fatty acid feedstocks. *Trans ASAE*; 46:945–55.

Canoira, L., Galean, J. G., Alacantara, R., Lapuerta, M. and Garcia, C. R. (2009). "Fatty Acid Methyl Esters (FAMES) from Castor oil: Production Process Assessment and Synergistic effects in its properties". *Renewable Energy*, pp. 1-10.

Cardone, M., Mazzoncini, M., Menini, S., Rocco, V., Senatore, A., Seggiani, M., Vitolo, S. (2003). *Biomass Bioenergy 25*, 623–636.

Carvalheiro, F., Duarte, L.C., Gírio, F.M. (2008). Hemicellulose biorefineries: a review on biomass pretreatments. *J Sci Ind Res* 67:849–864.

Chang, H.M., Liao, H.F., Lee, C.C., Shieh, C.J. (2005). Optimized synthesis of lipase-catalyzed biodiesel by Novozym 435. *J Chem Technol Biotechnol* 80(3):307–312.

Chen, J.W., Wu, W.T. (2003) Regeneration of immobilized Candida Antarctica lipase for transesterification. J Biosci Bioeng 92(2):231–237

Chevron (2006). Diesel fuels technical review. San Ramon, Cal.: Chevron Corp.

Christoffel, B.S., (2010). Evaluation of biodiesel from used cooking sunflower oil as substitute fuel: Tshwane University of Technology pp 1.

Couto, R. M., Fernandes, J., da Silva, M. D. R. G and Simoes, P. C. (2009). Supercritical fluid extraction of lipids from spent coffee grounds: *Journal of Supercritical Fluids,* vol. 51, pp. 159-166.

Cvengros, J. (1998). Acidity and corrosiveness of methyl esters of vegetable oils. *Fett/Lipid,* Vol. 100, No 2, pp. 41-44, ISSN 1521-4133.

Daglia, M., Racchi, M., Papetti, A., Lanni, C., Govoni, S., Gazzani, G (2004). In Vitro and ex Vivo Antihydroxyl Radical Activity of Green and Roasted Coffee. *J. Agric. Food Chem., 52,* 1700–1704.

De Oliveira, D., Di Luccio, M., Faccio, C., Dalla Rosa, C., Bender, J.P., Lipke, N., Menoncin, S., Amroginski, C., De Oliveira, J.V. (2004). Optimization of enzymatic production of biodiesel from castor oil in organic solvent medium. *Appl Biochem Biotechnol* 113–116:771 780.

De, A., Vieira, A.P., DaSilva, M.A.P., Langone, M.A.P. (2006). Biodiesel production via esterification reactions catalyzed by lipase: *Lat Am Appl Res*; 36:283 8.

Deligiannis, A., Papazafeiropoulou, A., Anastopoulos, G., Zannikos, F. (2011). Waste coffee grounds as an energy feedstock: *Proceedings of the 3rd International CEMEPE & SECOTOX Conference Skiathos, June 19-24, 2011, ISBN 978-960-6865-43-5.*

Demirbas A., (1997). Calculation of higher heating values of biomass fuels: *Fuel,* 76, 431 434.

Demirbas, A, Karsliog˘lu, S. (2007). Biodiesel production facilities from vegetable oils and animal fats: *Energy Sour*; 29:133–41.

Demirbas, A. (2003). Biodiesel fuels from vegetable oils via catalytic and non-catalytic supercritical alcohol transesterifications and other methods: a survey. *Energy Conv Mgmt*; 44:2093–2109.

Demirbas, A. (2005). Biodiesel production from vegetable oils via catalytic and non-catalytic supercritical methanol transes- terification methods. *Progress in Energy and Combustion science 31*: 466-487.

Demirbas, A. (2006). "Biodiesel Production via non-catalytic SCF method and biodiesel fuel characteristics": *Energy Conservation and Management,* vol. 47, pp. 2271-2282.

Deng, L., Xu, X.B., Haraldsson, G.G., Tan, T.W., Wang, F. (2005). Enzymatic production of alkyl esters through alcoholysis: A critical evaluation of lipases and alcohols. *J Am Chem Soc* 82(5):341–347

Dhabekar, A. and Chandak, A. (2010). Utilization of banana peels and beet waste foralcohol production. *Asiatic Journal of Biotechnology Resources Asiatic J. Biotech. Res. 2010; 01: 8 13.*

Dube, M.A., Tremblay, A.Y., Liu, J. (2007). Biodiesel production using a membrane reactor. *Bioresource Technology,* Vol. 98, pp. 639–647.

Encinar, J.M., Gonales, J.F., Rodriguez, J.J., Tejedor, A. (2002). Biodiesel fuels from vegetables Oils: Transesterification of Cynara cardunculus L. oils with ethanol. Energy and Fuels, 19: 443-450.

Fan, X.H., Wang, X., Chen, F., Geller, D.P., & Wan, P.J. (2008). Engine Performance Test of Cottonseed Oil Biodiesel: *The Open Energy and Fuels Journal*, 1, 40- 45.

Felizardo, P., Neiva Correia, M.J., Raposo, I., Mendes, J.F., Berkemeier, R., Moura Bordado, J., (2006). Biodiesel from waste frying oils: *Waste Manage*; 26:487–94.

Fellows, P. (1997). *Traditional Foods; Processing for profit. Intermediate Technology publication*: London.

Franca, A. S.; Oliveira, L. S.; and Ferreira, M. E. (2009). Kinetics and equilibrium studies of methylene blue adsorption by spent coffee grounds: *Desalination*, 249, 267– 272.

Franca, A.S.; Oliveira, L.S. (2009). Coffee processing solid wastes: In *Agricultural Wastes*, Ashworth, G.S., Azevedo, P., Eds.; Nova Science Publisher, Inc.: Hauppauge, NY, USA; Chapter 8, pp. 155–189.

Freedman, B., and M.O. Bagby, (1989). Heats of Combustion of Fatty Esters and Triglycerides,J. Am. *Oil Chem. Soc.* 66: 1601–1605.

Freedman, B., Butterfield, R.O., Pryde, E.H. (1986). Transesterification kinetics of soybean oil: *J Am Oil Chem Soc*; 63:1375–1380.

Freedman, B., Pryde, E.H. & Mounts, T.L. (1984). Variables affecting the yields of fatty esters from transesterified vegetable oils: *Journal of American Oil Chemists' Society,* Vol. 61, pp. 1638–1643.

Fukuda, H., Kondo, A., Noda, H. (2001). Biodiesel fuel production by transesterification of oils: *J Biosci Bioeng* 92(5):405–416.

Furuta, S., Matsuhasbi, H., Arata, K. (2004). Biodiesel fuel production with solid superacid catalysis in fixed bed reactor under atmospheric pressure: *Catal Commun*; 5:721–723.

Gathanju, d. (2010). Ethiopia Sets its Sights on Biodiesel. *Renewable Energy World.* http://www.renewableenergyworld.com/

Gemma, V., Mercedes, M. and Jose, A. (2004). Integrated Biodiesel Production: A comparison of different homogeneous catalysts systems: *Bioresource Technology*, vol. 92, pp. 297-305.

Ghadge, S.V. and Raheman, H. (2005). Biodiesel production from mahua (Madhuca indica) oil having high free fatty acids: *Biomass Bioenergy 28*, 601-605.

Giwa, S.O., Chuah, L.A., &Adam, N.M. (2010).Investigating "Egusi" *(Citrullus Colocynthis L.)* Seed Oil as Potential Biodiesel Feedstock*: Energies.* 3: 607-18

Goff, M.J., Bauer, N.S., Lopes, S., Sutterlin, W.R., Suppes, G.J. (2004). Acid-catalyzed alcoholysis of soybean oil: *J Am Oil Chem Soc*; 81:415–20.

Gressel, J. (2008). Transgenics are imperative for biofuel crops- Review. Plant Sci. 174: 246-263.

Haas, M.J., McAloon, A.J., Yee, W.C., Foglia, T.A. (2006). A Process Modelto Estimate Biodiesel Production Costs: *Bioresour Technol*, 97(4), 671- 678.

Hancsok, J., Bubalik, M., Beck, A., Baladincz. J. (2008). Development of multifunctional additives based on vegetable oils for high quality diesel and biodiesel: *Chem. Eng. Res. Des.* 86: 793–799.

Harrington, K.J., D'Arcy-Evans, C. (1985). Transesterification in situ of sunflower seed oil: *Ing Eng Chem Prod Res Dev*; 24:314–8.

Hass, M.J., Scott, K.M., Alleman, T.L., Mccormick, R.L. (2001). Engine performance of biodiesel fuel prepared from soybean soapstock: a high quality renewable fuel produced from a waste feedstock. *Energy Fuels*, 15: 1207-1212.

Helwani, Z., Othman, M.R., Aziz, N., Kim, J. and Fernando, W.J.N. (2009). Solid heterogeneous catalysts for transesterification of triglycerides with methanol: A review: *Appl. Catal.*, A 363, 1-10.

Hirata, S. and Berchmans H. J. (2007). "Biodiesel Production from crude Jatropha Curcas L. seed oil with high content of free fatty acids". *Bioresource Technology*, vol. 99, pp. 1716-1721.

Hoi, W. (2003). Potential of biomass utilization for energy in asia pacific sharing of specific experience of Philippine situation. Forest Research Institute. Kepong, Kuala Lumpur Malaysia. Usta, N. *Use of tobacco seed oil methyl ester in a turbocharged indirect injection diesel engine.* Biomass and bioenergy, (2005). 28: 77-86.

ICO, International Coffee Organization (2009). Total Coffee Production of Exporting Countries, London (UK) <www.ico.org/trade_statistics.asp> accessed.2013.

Kaieda, M., Samukawa, T., Matsumoto, T., Ban, M., Kondo, A., Shimada, Y. (1999). Biodiesel fuel production from plant oil catalysed by *Rhizopus oryzae* lipase in a water-containing system without an organic solvent: *J. Biosci. Bioeng.*88: 627-631.

Kerschbaum, S. and Rinke, G. (2004). Measurement of the temperature dependent viscosity of biodiesel fuels. *Fuel*; 83:287-291.

Kinast, J. A. (2003). Production of biodiesels from multiple feedstockes and properties of biodiesels and Biodiesel/Diesel blends. Final Reprort 1 in a series of 6. National Renewable Energy Laboratory, Golden, Colorado: 80401-3393.

Kingwell, R., and B. Plunkett (2006). Economics of On-Farm Biofuel Production: Invited Paper at Bioenergy and Biofuels.

Knothe, G. (2002). "Structure indices in FA chemistry. How relevant is the iodine value?" *Journal of the American Oil Chemists' Society*, vol. 79, no. 9, pp. 847–854.

Knothe, G. (2005). "Dependence of Biodiesel Fuel Properties on the Structure of Fatty Acid Alkyl Ester": *Fuel Processing Technology*, vol. 86, pp.1059-1070.

Knothe, G. (2006). Analysing Biodiesel: Standards and other methods". JAOCS, vol. 83, pp. 823-833.

Knothe, G. (2007). Some aspects of biodiesel oxidative stability: *Fuel Process Technology*; 88:669–77.

Knothe, G. (2008). "Designer" biodiesel: optimizing fatty ester composition to improve fuel properties. Energ. Fuel 22: 1358–1364.

Knothe, G. and Dunn, R.O. (2001). In: F.D. Gunstone, R.J. Hamilton (Eds.), Oleochemical Manufacture and Applications, Sheffield Academic Press, Sheffield, UK, pp. 106.

Knothe, G. and Steidley, R., K. (2005). "Kinematic Viscosity of Biodiesel Fuel components and related compounds: Influence of compound structure and comparison to petro-diesel fuel Components": *Fuel*, vol. 84, pp. 1059-1065.

Knothe, G., Van Gerpen, J., and Krahl, J. (2005). The Biodiesel Handbook: Champaign, IL; AOCS Press.

Kochhar,S.L(1998).Economic Botany in the Tropics. 2nd edition: *Macmillan India Ltd.*

Kondamudi, N., Mohapatra, S. K., Misra, M. (2008). Spent coffee grounds as a versatile source of green energy: *Journal of Agricultural and Food Chemistry*, 56; 11757–11760.

Korytkowska, A., Barszczewska-Rybarek, I., Gibas, M. (2001). Side-reactions in the transesterification of oligoethylene glycols by methacrylates: *Des Monom Polym*; 4:27–37.

Kosmehl, S. O., Heinrich, H. (1997). The Automotive Industry's View on the Standards for Plant Oil-Based Fuels, In: *Plant Oils as Fuels. Present State of Science and Future Developments*, N. Martini & J. Schell (Eds.), 18-28, ISBN 3540647546 9783540647546, Berlin, Germany.

Kulkarni, M.G. (2006). Waste cooking oil – an economical source for biodiesel: a review: *Ind Eng Chem Res* vol. 45, pp. 2901–13.

Kusdiana, D., Saka, S. (2001). Kinetics of transesterification in rapeseed oil to biodiesel fuels as treated in supercritical methanol. Fuel; 80:693–8.

Lague, C., Lo, K., Staley, L. (1987). Waste Vegetable Oil as a Diesel Fuel Extender: *Canadian Agric. Eng.* 29: 27-32.

Lai, C.C., Zullaikah, S., Vali, S.R., Ju, Y.H. (2005). Lipase-catalyzed production of biodiesel from rice bran oil: *J Chem Technol Biotechnol* 80(3):331–337.

Lai, O.M., Ghazali, H.M., Chong, C.L. (1999). Use of enzymatic transesterified palm stearin–sunflower oil blends in the preparation of table margarine formulation: *Food Chem*; 64:83–8.

Lamare, M. and Wing, S. (2001). Calorific content of New Zealand marine macrophytes: *New Zealand Journal of Marine and Freshwater Research*; 35:335-341.

Lee, Y., Park, S.H., Lim, I.T., Han, K., Lee, S.Y. (2000). Preparation of alkyl (R)-(2)-3-hydroxybutyrate by acidic alcoholysis of poly-(R)-(2)-3-hydroxybutyrate: *Enzyme Microb Technol*; 27:33–6.

Leifa, F., Pandey, A.,&Soccol, C. R. (2000). Solid state cultivation—An efficient method to use toxic agro-industrial residues: *Journal of Basic Microbiology*, 40, 187–197.

Leung, D.Y.C., Guo, Y. (2006). Transesterification of neat and used frying oil:Optimization for biodiesel production: *Fuel Process Technol*; 87:883–90

Linko, Y.Y., Lamsa, M., Wu, X., Uosukainen, W., Sappala, J., Linko, P. (1998). Biodegradable products by lipase biocatalysis: *J Biotechnol*; 66:41–50.

Liu, X., Xiong, X., Liu, C., Liu, D., Wu, A., Hu, Q., Chunlin, L. (2010). Preparation of biodiesel by transesterification of rapeseed oil with methanol using solid base catalyst Calcined K. *Journal of the American Oil Chemists' Society,* 87, 817-823.

Liu, Y., Lotero, E., Goodwin, J.G. (2006). Effect of water on sulfuric acid catalyzed esterification: *J Molec Catal A: Chem*; 245:132–40.

Loh, S. K., Chew, S. M., Choo, Y. M. (2006). Oxidative stability and storage behavior of fatty acid methyl esters derived from used palm oil: *J. Am. Oil Chem. Soc.* 83: 947–952.

Lopez, D.E., Goodwin, J.G., Bruce, D.A., Lotero, E. (2005). Transesterification of triacetin with methanol on solid acid and base catalysts: *Appl Catal A: Gen*; 295:97–105.

Lotero, E., Goodwin, J.G., Bruce, D.A., Suwannakarn, K., Liu, Y., Lopez, D.E. (2006). The catalysis of biodiesel synthesis: *Catalysis*; 19:41–83.

Ma, F., Clements, L.D., Hanna, M.A. (1998). The effect of catalyst, free fatty acids, and water on transesterification o beef tallow: *Trans ASAE*; 41:1261–4.

Ma, F., Hanna, M. A., (1999). Biodiesel production: A review: *Bioresour. Tech.*, 70 (1), 1-15

Maceiras, R., Vega, M., Costa, C., Ramos, P., Márquez, M.C. (2009). Effect of methanol content on enzymatic production of biodiesel from waste frying oil: *Fuel 88*, 2130–2134.

Mallinckrodt Baker Inc., (2009). Methyl Alcohol Material Safety Data Sheet.

Marchetti, J.M., Miguel, V.U., Errazu, A.F. (2005). Homogeneous catalyst for the esterification of high free fatty acid content oils: In: 13- Congresso Brasileiro de Catálise 3-Congresso de Catálise do Mercosul.

May, C.Y. (2004). Transesterification of palm oil: effect of reaction parameters. *J Oil Palm Res*; 16:1–11.

Meher, L. C.; Vidya, D. Sagar and Naik, S. N. (2004). Technical aspects of biodiesel production by transterification - a review: *Renewable and Sustainable Energy Reviews 10*: 248-268.

Meher, L.C., Dharmagadda, V.S.S., Naik, S.N. (2006). Optimization of alkali-catalyzed transesterification of Pongamia pinnata oil for production of biodiesel: *Bioresource. Technol.* 97: 1392-1397.

Meher, L.C., Naik, S.N., Das, L. (2004). Methanolysis of Pongamia pinnata (Karanja) oil for production of biodiesel: *Bioresource Technology*. 97:1392-1397.

Meher, L.C., Sagar, D.V., Naik, S.N. (2006). Technical aspects of biodiesel production by transesterification-a review: Renew. *Sust. Energ. Rev.*, 10 (3), 248-268.

Meneghetti, S.M., Plentz, M.R., Meneghetti, C.R., Wolf, E.C., Silva, G.E., S., Lima, L.d.S., Tatiana, M., Serra, F.C., Lenise, G., de Oliveira, (2006). "Biodiesel from Castor Oil: A Comparison of Ethanolysis versus Methanolysis". *Energy and Fuels*, vol. 20, pp. 2262-2265.

Mesfin Kinfu (2008). M.Sc. thesis: Investigation of Alternative Locally Available Feedstock Sources for Biodiesel Production in Ethiopia: Addis Ababa University.

Meskir Tesfaye (2007). Bio-fuels in Ethiopia: *Eastern and Southern Africa Regional Workshop on Bio-fuels*

Miao, X., Wu, Q. (2006). Biodiesel production from heterotrophic microalgal oil: *Bioresource Technology*; 97:841–6.

Midwest Biofuels (1994). Biodiesel Cetane Number Engine Testing Comparison to Calculated Cetane Index Number: Midwest Biofuels, U.S.A.

Ministry of Mines and Energy of Ethiopia (2011). Energy Policy of Ethiopia: Japan International Cooperation Agency (JICA) Tokyo international center.

Mittelbach, M. (1996). Diesel Derived from Vegetable Oils, VI: Specifications and Quality Control of Biodiesel; Bioresource Technology; 56:7–11.

Mittelbach, M. and Gangl, S. (2001). Long Storage Stability of Biodiesel made from Rapeseed and Used Frying Oil; *J of American Oil Ch. Soc.*; 78:573–577.

Mohammed, M. (2011). Mathematical Modeling of a Two-Phase Bubble-Column Reactor for Biodiesel Production from Alternative Feedstocks: Chemical Engineering, Drexel University.

Moser, B.R. (2009). Biodiesel production, properties, and feedstocks: *In Vitro Cell.Dev.Biol. — Plant.* 45: 229–266.

Moser, B.R.; Erhan, S.Z. (2008). Branched chain derivatives of alkyl oleates: tribological, rheological, oxidation, and low temperature properties: *Fuel* 87: 2253–2257.

Mussatto, S.I., Machado, E.M.S., Martins, S., Teixeira, J.A. (2011). Production, composition, and application of coffee and its industrial residues: *Food Bioprocess Technol. 4*, 661–672.

Mussatto, S.I., Teixeira, J.A. (2010). Lignocellulose as raw material in fermentation processes. In: Méndez-VilasA(ed) Current research, technology and education topics in *applied microbiology and microbial biotechnology*, vol 2. Formatex Research Center, Badajoz, Spain, pp 897–907

National Biodiesel Board (2010). Biodiesel 101: *Biodiesel: http://www.biodiesel.org/resources/*

Nelson, L.A., Foglia, T.A., Marmer, W.N. (1996). Lipase-catalyzed Production of biodiesel: *J.Am.Oil Chem..Soc.* 73; 1191–1195.

Nkolika, I.C., Obiajulu, B.O.C., Uwaoma, O. (2012). Comparative Study of the Physicochemical Characterization of Some Oils as Potential Feedstock for Biodiesel Production: *ISRN Renewable Energy* Volume 2012 (2012), Article ID 621518, 5 pages

Noureddini, H., Gao, X., Philkana, R.S. (2005). Immobilized Pseudomonas cepacia lipase for biodiesel fuel production from soybean oil: *Bioresour Technol* 96(7):769–777.

Noureddini, H., Harkey, D., Medikonduru, V.A. (1998). Continuous process for the conversion of vegetable oil into methyl esters of fatty acids: *J. AOCS*, 75: 1775-1783.

Nutawan, Y., Phattayawadee, P., Pattranit, T., Mohammad, N. (2010). Bioethanol Production from Rice Straw: *Energy Research J.* 1: 26-31.

Olaniyan, A.M., Oje, K. (2007). Quality characteristics of shea butter recovered form shea kernel through dry extraction process: *J. Food Sci. Technol.*, 44(4): 404-407.

Omolara, O.O., Dosumu, O.O. (2009). "Preliminary Studies on the effect of processing methods on the quality of three commonly consumed marine fishes in Nigeria:" *Biokemistri*, vol. 21, no. 1, pp. 1–7.

Parr (1987). Operating Instructions for the 1241 Oxygen Bomb Calorimeter. Parr Instrument Company. Molene, Illinois. USA.

Pate, B. and Gami, B. (2012). Biomass Characterization and its Use as Solid Fuel for Combustion: *Iranica Journal of Energy & Environment* 3 (2): 123-128.

Peterson, C.L., Feldman, M., Korus, R., Auld, D.L. (1991). Batch type transesterification process for winter rape oil: *Appl Eng Agric*; 7:711–716.

Poskart, M. and Szecówka, L. (2008). Energy utilization of wood waste: *Archivum Combustionis*, 28, 57-66.

Pramanik, K. (2003). Properties and use of jatropha curcas oil and diesel fuel blends in compression ignition engine. *Renewable Energy* 28: 239-248.

Prankl, H. and Worgetter, M. (1996). Influence of the Iodine Number of Biodiesel to the Engine Performance: Proceedings of the Alternative Energy Conference, American Society of Agricultural Engineers.

Quaak, P., Knoef, H., Stassen, H. (1999). Energy from biomass (A review of combustion and gasification techniques): New York, USA, 2-7.

Quevedo-Hidalgo, B., Monsalve-Marín, F., Narváez-Rincón, P.C., Pedroza-Rodríguez, A.M., Velásquez-Lozano, M.E. (2013). Ethanol production by Saccharomyces cerevisiae using lignocellulosic hydrolysate from Chrysanthemum waste degradation: *World J Microbiol Biotechnol*, 459-466.

Ramadhas, A.S.; Jayaraj, S., Muraleedharan, C. (2005). Characterization and the effects of using rubber seed oil as fuel in the compression ignition engines: *Renewable Energy* vol. 30 no.5 pp. 795-803.

Rashid, U., Anwar, F., &Arif, M. (2009).Optimization of base catalytic methanolysis of sunflower (Helianthus annuus) seed oil for biodiesel production by using response surface methodology. *Ind. Eng. Chem. Res*. 48: 1719-26.

Refaat, A. A., (2010). Different techniques for the production of biodiesel from waste vegetable oil: *Int. J. Environ. Sci. Tech*., 7 (1), 183-213.

Rossel, J.B. (1987). Classical analysis of oils and fats; In: Hamilton RJ, Rossel JB (Eds): Analysis of oils and fats. Elsevier Applied Science, London and New York, pp. 1-90.

Royon, D., Daz, M., Ellenrieder, G., Locatelli, S. (2007). Enzymatic production of biodiesel from cotton seed oil using t-butanol as a solvent: *Bioresour Technol*; 98:648–653.

Sahoo, P.K., Das, L.M., Babu, M.K.G., Naik, S.N. (2007).Biodiesel development from high acid value polanga seed oil and performance evaluation in a CI engine: *Fuel*, 86, 448-454.

Sánchez, A., Maceiras, R., Cancela, A., Rodríguez, M. (2012). Influence of n-Hexane on in Situ Transesterification of Marine Macroalgae: *Energies* 5, 243-257.

Sanford, S.D., et al., (2009). Feedstock and Biodiesel Characteristics Report: *Renewable Energy Group*, Inc., www.regfuel.com.

Santori, G., Nicola, G. D., Moglie, M., Polonara, F. (2012). *A review analyzing the industrial* biodiesel production practice starting from vegetable oil refining: *Applied Energy*, 92, 109-132.

Sarma, A.K., Konwer, D., Bordoloi, P. K. (2005). A Comprehensive analysis of fuel properties of biodiesel from Koroch seed oil: *Energ. Fuel* 19: 656–657.

Sawyer, E. (2007). The Feasibility of On-farm Biodiesel Production: Department of Agricultural Economics, Saskatchewan University.

Schafer, A. D. Naber and Gairing, M. (1997).Biodiesel als alternativer Kraftstoff für Mercedes-Benz-Dieselmotoren: *Mineralöltechnik,* Vol. 43, pp. 1-32, ISSN: 0307-6490.

Scherpenzeel, J. (2000). The use of biofuel in diesel engines: tests on a Fiat 160/90 DT tractor. European Energy Crops InterNetwork.

Sendzikiene, E., Makareviciene, V., Janulis, P., Kitrys, S. (2004). *Eur.* Kinetics of free fatty acids esterification with methanol in the production of biodiesel fuel: *J. Lipid Sci. Technol, 106,* 831–836.

Shah, S., Gupta, M.N. (2007).Lipase catalyzed preparation of biodiesel from Jatropha oil in a solvent free system: *Process Biochem*; 42:409–14.

Sharma, Y. C., Singh, B., Upadhyay, S.N. (2008).Advancements in development and characterization of biodiesel: a review. Fuel 87: 2355–2373.

Sharp, C.A., Howell, S. Jobe, J. (2000).The Effect of Biodiesel Fuels on Transient Emissions from Modern Diesel Engines, Part I Regulated Emissions and Performance, *SAE* 2000-01-1967.

Shimada, Y., Watanabe, Y., Sugihara, A., Tominaga, Y., (2002). Enzymatic alcoholysis for biodiesel fuel production and application of the reaction to oil processing. *J. Molecul. Catal. B: Enzymatic*, 17 (3-5), 133-142.

Silva, M.A., Nebra, S.A., Silva, M.J.M., Sanchez, C. G. (1998).The use of biomass residues in the Brazilian soluble coffee industry: *Biomass and Bioenergy* 14, 457–467.

Simões, J., Madureira, P., Nunes, F.M., Domingues, M.R., Vilanova, M., Coimbra, M.A. (2009). Immunostimulatory properties of coffee mannans: Molecular Nutrition & Food Research, 53, 1036–1043.

Singh, A.B., He, C., Thompson, J., Van Gerpen, J., (2006).Process optimization of biodiesel production using different alkaline catalysts: *Applied Engineering in Agriculture,* Vol. 22, No.4, pp. 597-600.

Sivaramakrishnan, K., Ravikumar, P., (2012). Determination of cetane number of biodiesel and its influence on physical properties: *ARPN Journal of Engineering and Applied Sciences*: VOL. 7, NO. 2.

Skagerlind, P., Jansson, M., Bergenstahl, B., Hult, K. (1995).Binding of Rhizomucor miehei lipase to emulsion interfaces and its interference with surfactants: Colloids Surf B: Bioint; 4:129–35.

Srivastava, A., Prasad, R. (2000). Triglycerides-based diesel fuels. Renew Sust Energy Rev; 4:111–33.

Stavarache, C., Vinatoru, M., Nishimura, R., Maeda, Y. (2005). Fatty Acids Methyl Esters from Vegetable Oil by Means of Ultrasonic Energy: *Ultrason. Sonochem.* 12, 367-372.

Stern, R., Hillion, G. (1990). Purification of esters: Eur Pat Appl EP 356317. (Cl. C07C67/ 56) [c.f. Chem Abstr 113: 58504k].

Strong, C., Erickson, C., Shukla, D. (2004). Evaluation of Biodiesel Fuel: Literature Review. Western Transportation Institute.

The African Commodity Report (2012). Delmas Com-Watch – issue 9.

Tokimoto, T., Kawasaki, N., Nakamura, T., Akutagawa, J.,&Tanada, S. (2005). Removal of lead ions in drinking water by coffee grounds as vegetable biomass. *Journal of Colloid and Interface Science* 281, 56–61.

U.S. Environmental Protection Agency (EPA) (2008).Starbucks Protects Environment by Offering Grounds for Your Garden: Resource Conservation Challenge (RCC); Eighteen DEC. Six Mar 2008 <http://www.epa.gov/epaoswer/osw/conserve/2004news/ 04-star.htm>.

Unesco. (1988). Engineering schools and endogenous technology development: Paris.

Usta, N. (2005).Use of tobacco seed oil methyl ester in a turbocharged indirect injection diesel engine*: Biomass and bioenergy.*28: 77-86.

Van Gerpen, J. et al. (1996). Determining the Influence of Contaminants on Biodiesel Properties; Iowa State University for The Iowa Soybean Promotion Board.

Van Gerpen, J., Shanks, B., Pruszko, R., Clements, D., Knothe, G. (2004). Biodiesel Analytical Methods: National Renewable Energy Laboratory: 1617 Cole Boulevard, Colorado. U.S.

Van Gerpen, J.V. (2005). Biodiesel Processing and Production: *Fuel processing Technology*, vol. 86, no.10, pp. 1097-1107.

Varghaa, V., Truterb, P. (2005).Biodegradable polymers by reactive blending transesterification of thermoplastic starch with poly (vinyl acetate) and poly (vinyl acetate-co-butyl acrylate: *Eur Polymer J*; 41:715–26.

Veljkovi´c, V. B., Laki´cevi´c, S. H., Stamenkovi´c, O. S., Todorovi´c, Z. B., Lazi´c, M. L. (2006). Biodiesel production from tobacco (*Nicotiana tabacum L.*) seed oil with a high content of free fatty acids: *Fuel* 85: 2671–2675.

Vicente, G., Martínez, M., Aracil, J. (2004). Integrated biodiesel production: a comparison of different homogeneous catalysts systems: *Biores Technol.*, 92: 297-305.

W/Giorgis, A. (2004). Rural Development Strategy for Ethiopia: *Symposium on Renewable Energies in Ethiopia* (p. 30). Addis Ababa: Ethiopian Rural Energy Development and Promotion Center,GTZ.

Watanabe, Y., Shimada, Y., Sugihara, A., Tominaga, T. (2002).Conversion of degummed soybean oil to biodiesel fuel with immobilized Candida antarctica lipase: *J Mol Catal B: Enzym*; 17:151–155.

Weast, R.C., Astle, M.J., Beyer, W.H. (1985–1986).Handbook of Chemistry and Physics, 66th edn., CRC Press, Boca Raton, FL, pp. D-272–D-278.

Wright, H.J., Segur, J.B., Clark, H.V., Coburn, S.K., Langdom, E.E., DuPuis, R.N. (1944). A report on ester interchange: *Oil and Soap*; 21:145–8.

www.worldmapper.org, "Worldmapper coffee production," ed, 2006.

Xu, Y.Y., Du, W., Zeng, J., Liu, D.H. (2004). Conversion of soybean oil to biodiesel fuel using lipozyme TL 1M in a solvent-free medium: *Biocatal Biotransformation* 22(1):45–48.

Yamada, R., Taniguchi, N., Tanaka, T., Ogino, C., Fukuda, H., and Kondo, A., 2011. Direct ethanol production from cellulosic materials using a diploid strain of *Saccharomyces cerevisiae* with optimized cellulase expression: Biotechnology for Biofuels 2011, 4:8

Yanagimoto, K., Ochi, H., Lee, K.G., Shibamoto, T. (2004).Antioxidative Activities of Fractions Obtained from Brewed Coffee: *J. Agric. Food Chem.52*, 592–596.

Yen, W.J., Wang, B.S., Chang, L.W., Duh, P.D. (2005).Antioxidant properties of roasted coffee residues: *Journal of Agricultural and Food Chemistry* 53, 2658–2663.

Yong, O.Y. and Salimon, J. 2006. Characteristics of *Elateriospermum tapos* seed oil as a new source of oilseed: *Ind. Crops Prod. 24*, 146–151.

Zhang, Y., Dube, M.A., McLean, D.D., Kates, M. (2003).Biodiesel production from waste cooking oil: 1. Process design and technological assessment: *Bioresour Technol*; 89:1–16.

Zheng, D. and Hanna, M. (1995).Preparation and Properties of Methyl Esters of Beef Tallow: Department of Biological Systems Engineering, University of Nebraska-Lincoln, USA.

APPENDICES

Appendix 1: ASTM D6751-09 Standard Specification for Biodiesel Fuel (B100)

Property	ASTM Method	Limits	Units
Flashpoint	D93	93.0 Min	°C
Water & Sediment	D2709	0.050 Max	% Vol.
Kinematic Viscosity, 40°C	D445	1.9-6.0	mm^2/Sec.
Sulfated Ash	D874	0.020 Max	% Mass.
Sulfur			
S 15 Grade	D 5453	0.0015 max	% mass(ppm)
S 500 Grade	D 5453	0.05 max	% mass(ppm)
Copper Strip corrosion	D130	No. 3 Max	
Cetane	D613	47 Min	
Cloud point	D2500	Report	°C
Carbon Residue	D4530	0.050 Max	% Mass
Acid Number	D664	0.50 Max	mg KOH/g
Free Glycerine	D6584	0.020 Max	% Mass.
Total Glycerine	D6584	0.24 Max	% Mass.
*Calcium & Magnesium combined	EN 14538	5 Max	ppm (µg/g)
*Alcohol Control (one of the following must be met)			
1. Methanol Content	EN14110	0.2 Max	%volume
2. Flash Point	D93	130 Min	°C
*Phosphorus Content	D 4951	0.001 Max	% mass
Distillation Temperature Atmospheric	D 1160	360 Max	°C

equivalent temperature, 90 % recovered			
Sodium/Potassium combined	EN 14538	5 Max	Ppm
Oxidation Stability	EN 14112	3 minimum	Hours
Cold Soak Filtration	Annex to D6751	360 Max	seconds
For use in temperatures below -12°C	Annex to D6751	360 Max	seconds

[a]For the table: min refers to minimum and max refers to maximum

[b]The limits are for Grade S15 and Grade S500 biodiesel, with S15 and S500 referring to maximum allowable sulfur content (ppm)

Source: ASTM (2009)

Appendix 2: EN 14214 biodiesel fuel standard

Property	Test method	Limits	Unit
Ester content	EN 14103	96.5 min	% (mol/mol)
Density; 15°C	EN ISO 3675 EN ISO 12185	860 – 900	kg/m^3
Kinematic viscosity, 40°C	EN ISO 3104, ISO 3105	3.5- 5.0	mm^2/s
Flash point	EN ISO 3679	120 min	°C
Sulfur content	EN ISO 20846, EN ISO 20884	10.0 max	mg/kg
Carbon residue (10% dist. residue)	EN ISO 10370	0.30 max	% (mol/mol)
Cetane number	EN ISO 5165	51 min	
Sulfated ash	ISO 3987	0.02 max	% (mol/mol)
Water content	EN ISO 12937	50 max	mg/kg
Total contamination	EN 12662	24 max	mg/kg
Copper strip corrosion (3h, 50°C)	EN ISO 2160	1	Degree of corrosion
Oxidative stability, 110°C	EN 14112	6.0 min	H
Acid value	EN 14104	0.5 max	mg KOH/g

Iodine value	EN 14111	120 max	g I_2/100g
Linolenic acid content	EN 14103	12.0 max	% (mol/mol)
Polyunsaturated (≥4 double bonds) methyl esters	EN 14103	1 max	% (mol/mol)
Methanol content	EN 14110	0.20 max	% (mol/mol)
MAG content	EN 14105	0.80max	% (mol/mol)
DAG content	EN 14105	0.20 max	% (mol/mol)
TAG content	EN 14105	0.20 max	% (mol/mol)
Free glycerol	EN 14105, EN 14106	0.020 max	% (mol/mol)
Total glycerol	EN 14105	0.25 max	% (mol/mol)
Group I metals (Na, K)	EN 14108, EN 14109	5.0 max	mg/kg
Group II metals (Ca, Mg)	EN 14538	5.0 max	mg/kg
Phosphorus content	EN 14107	10.0 max	mg/kg

Source: (Moser, 2009).

Appendix 3: Fatty acid composition of waste coffee residue oil methyl ester

Fatty acid	Chemical structure		Weight %
Palmitic	**C 16**	CH3(CH2)14COOH	37.6
Stearic	**C 18**	CH3(CH2)16COOH	7.6
Oleic	**C 18:1**	CH3(CH2)7CH=CH(CH2)7COOH	12.7
Linoleic	**C 18:2**	CH3(CH2)3(CH2CH=CH)2(CH2)7COOH	39.8
*	-		0.3*
*	-		2.0*

*unidentified

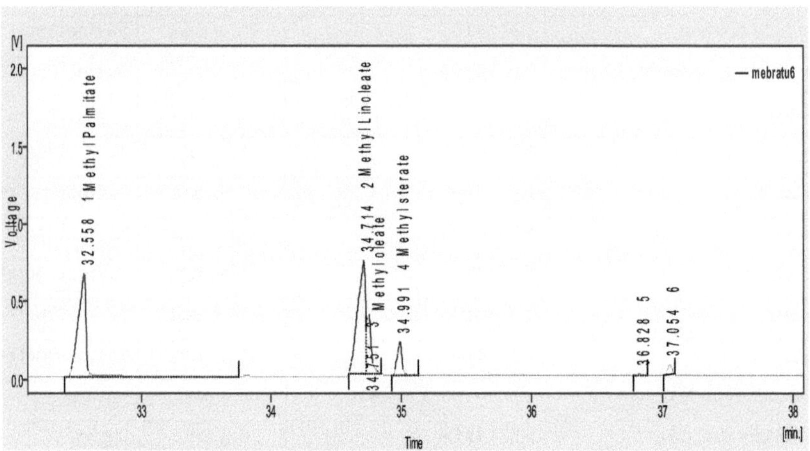

Appendix 5: Distillation Characteristics of WCR Methyl Ester

Percentage Recovery (cm3)	Temperature (^{o}C)
IBP	314.5
5%	324.5
10%	327
15%	328.5
20%	327.5
25%	326.5
30%	327.5
35%	327.5
40%	328.5
45%	329.5
50%	330
55%	331
60%	332
65%	332.5
70%	334.5

	75%		336

75%		336
80%		340
85%		343.5
90%		349
95%		360
FBP		362.5
% Total Recovery		97.77

Appendix 6: Bioethanol yield from spent of WCR

Types of hydrolysis	Amount collected (g)	Concentration	Gram of ethanol	Yield of ethanol (%)	
				w/w	v/v
Distilled water	14.053	8.71	1.224	4.9	6.2
1M	11.774	13.87	1.633	6.53	8.27
2M	12.345	4.45	0.55	2.2	2.8
3M	10.421	4.83	0.5	2	2.5

Appendix 7: Absorbance reading of spent of WCR bioethanol with distilled water and 1 molar H_2SO_4

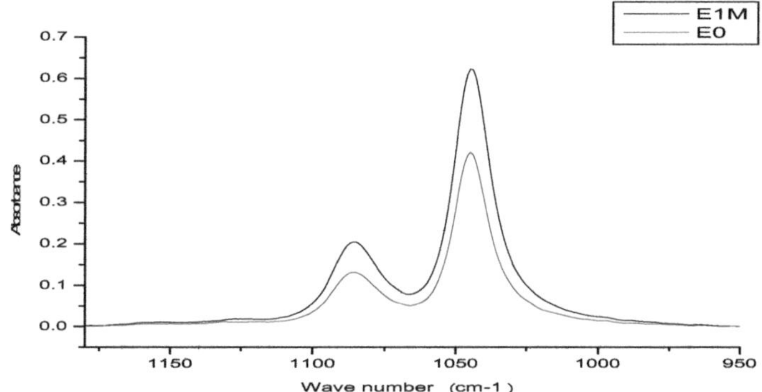

Run	Proximate analysis			
	Moisture	Volatile matter	Ash	Fixed carbon
1	4.43	75.73	2.04	18.01
2	4.57	75.91	2.12	17.40
3	4.5	75.97	1.96	17.35
Average	**4.5**	**75.87**	**2.04**	**17.59**

Appendix 9: calculations for esterification and transesterification

Appendix 9.1: calculation for Esterification

$$Kg \text{ (methanol)} = kg \text{ oil} \times \frac{\%FFA}{100} \times \text{molar ratio} \times \frac{\text{molecular weight methanol}}{\text{molecular weight FFAs}}$$

$$Kg \text{ (methanol)} = 0.5 \text{ kg oil} \times \frac{\%7.33}{100} \times 20 \times \frac{32 \frac{g}{mol}}{282.5 \frac{g}{mol}} = 0.083 \text{ kg}$$

$$L \text{ (methanol)} = \frac{0.083 \text{ kg}}{0.7914 \text{ kg/L}} = 0.10488 \text{ L} = 104.88 \text{ mL}$$

$$Kg \text{ (H2SO4)} = Kg \text{ oil} \times \frac{\%FFA}{100} \times 0.1$$

Appendix 9.2: calculation for Transesterification

$$\boldsymbol{Kg \ (methanol) = kg \ oil \times \frac{\%FFA}{100} \times molar \ ratio \times \frac{32 \ g/mol}{molecular \ weight \ FFAs}}$$

$$Kg \text{ (methanol)} = 0.457 \times \frac{7.33}{100} \times 6 \times \frac{\text{molecular weight methanol}}{885.46 \frac{g}{mol}} = 0.0073 \text{ kg}$$

$$L \text{ (methanol)} = \frac{0.0073 \text{ kg}}{0.7914 \text{ kg/L}} = 0.00922 \text{ L} = 9.22 \text{ mL}$$

TOMOCA PLC for waste coffee grounds sample

sample sun drying

Soxhlet Extraction

Adiabatic oxygen bomb for calorific value